TO QUEUE OR NOT TO QUEUE

INTERNATIONAL SERIES IN
OPERATIONS RESEARCH & MANAGEMENT SCIENCE
Frederick S. Hillier, Series Editor **Stanford University**

Weyant, J. / *ENERGY AND ENVIRONMENTAL POLICY MODELING*
Shanthikumar, J.G. & Sumita, U. / *APPLIED PROBABILITY AND STOCHASTIC PROCESSES*
Liu, B. & Esogbue, A.O. / *DECISION CRITERIA AND OPTIMAL INVENTORY PROCESSES*
Gal, T., Stewart, T.J., Hanne, T. / *MULTICRITERIA DECISION MAKING: Advances in MCDM Models, Algorithms, Theory, and Applications*
Fox, B.L. / *STRATEGIES FOR QUASI-MONTE CARLO*
Hall, R.W. / *HANDBOOK OF TRANSPORTATION SCIENCE*
Grassman, W.K. / *COMPUTATIONAL PROBABILITY*
Pomerol, J-C. & Barba-Romero, S. / *MULTICRITERION DECISION IN MANAGEMENT*
Axsäter, S. / *INVENTORY CONTROL*
Wolkowicz, H., Saigal, R., & Vandenberghe, L. / *HANDBOOK OF SEMI-DEFINITE PROGRAMMING: Theory, Algorithms, and Applications*
Hobbs, B.F. & Meier, P. / *ENERGY DECISIONS AND THE ENVIRONMENT: A Guide to the Use of Multicriteria Methods*
Dar-El, E. / *HUMAN LEARNING: From Learning Curves to Learning Organizations*
Armstrong, J.S. / *PRINCIPLES OF FORECASTING: A Handbook for Researchers and Practitioners*
Balsamo, S., Personé, V., & Onvural, R./ *ANALYSIS OF QUEUEING NETWORKS WITH BLOCKING*
Bouyssou, D. et al. / *EVALUATION AND DECISION MODELS: A Critical Perspective*
Hanne, T. / *INTELLIGENT STRATEGIES FOR META MULTIPLE CRITERIA DECISION MAKING*
Saaty, T. & Vargas, L. / *MODELS, METHODS, CONCEPTS and APPLICATIONS OF THE ANALYTIC HIERARCHY PROCESS*
Chatterjee, K. & Samuelson, W. / *GAME THEORY AND BUSINESS APPLICATIONS*
Hobbs, B. et al. / *THE NEXT GENERATION OF ELECTRIC POWER UNIT COMMITMENT MODELS*
Vanderbei, R.J. / *LINEAR PROGRAMMING: Foundations and Extensions, 2nd Ed.*
Kimms, A. / *MATHEMATICAL PROGRAMMING AND FINANCIAL OBJECTIVES FOR SCHEDULING PROJECTS*
Baptiste, P., Le Pape, C. & Nuijten, W. / *CONSTRAINT-BASED SCHEDULING*
Feinberg, E. & Shwartz, A. / *HANDBOOK OF MARKOV DECISION PROCESSES: Methods and Applications*
Ramík, J. & Vlach, M. / *GENERALIZED CONCAVITY IN FUZZY OPTIMIZATION AND DECISION ANALYSIS*
Song, J. & Yao, D. / *SUPPLY CHAIN STRUCTURES: Coordination, Information and Optimization*
Kozan, E. & Ohuchi, A. / *OPERATIONS RESEARCH/ MANAGEMENT SCIENCE AT WORK*
Bouyssou et al. / *AIDING DECISIONS WITH MULTIPLE CRITERIA: Essays in Honor of Bernard Roy*
Cox, Louis Anthony, Jr. / *RISK ANALYSIS: Foundations, Models and Methods*
Dror, M., L'Ecuyer, P. & Szidarovszky, F. / *MODELING UNCERTAINTY: An Examination of Stochastic Theory, Methods, and Applications*
Dokuchaev, N. / *DYNAMIC PORTFOLIO STRATEGIES: Quantitative Methods and Empirical Rules for Incomplete Information*
Sarker, R., Mohammadian, M. & Yao, X. / *EVOLUTIONARY OPTIMIZATION*
Demeulemeester, R. & Herroelen, W. / *PROJECT SCHEDULING: A Research Handbook*
Gazis, D.C. / *TRAFFIC THEORY*
Zhu, J. / *QUANTITATIVE MODELS FOR PERFORMANCE EVALUATION AND BENCHMARKING*
Ehrgott, M. & Gandibleux, X. / *MULTIPLE CRITERIA OPTIMIZATION: State of the Art Annotated Bibliographical Surveys*
Bienstock, D. / *Potential Function Methods for Approx. Solving Linear Programming Problems*
Matsatsinis, N.F. & Siskos, Y. / *INTELLIGENT SUPPORT SYSTEMS FOR MARKETING DECISIONS*
Alpern, S. & Gal, S. / *THE THEORY OF SEARCH GAMES AND RENDEZVOUS*
Hall, R.W./*HANDBOOK OF TRANSPORTATION SCIENCE - 2nd Ed.*
Glover, F. & Kochenberger, G.A. / *HANDBOOK OF METAHEURISTICS*
Graves, S.B. & Ringuest, J.L. / *MODELS AND METHODS FOR PROJECT SELECTION: Concepts from Management Science, Finance and Information Technology*

TO QUEUE OR NOT TO QUEUE: EQUILIBRIUM BEHAVIOR IN QUEUEING SYSTEMS

REFAEL HASSIN
Department of Statistics and Operations Research
Tel Aviv University
Tel Aviv 69978, Israel
hassin@post.tau.ac.il

MOSHE HAVIV
Department of Statistics
The Hebrew University
Jerusalem 91905, Israel,
and
Econometrics and Business Statistics
The University of Sydney
Sydney NSW 2006, Australia
haviv@mscc.huji.ac.il.

Kluwer Academic Publishers
Boston/Dordrecht/London

Distributors for North, Central and South America:
Kluwer Academic Publishers
101 Philip Drive
Assinippi Park
Norwell, Massachusetts 02061 USA
Telephone (781) 871-6600
Fax (781) 871-9045
E-Mail: kluwer@wkap.com

Distributors for all other countries:
Kluwer Academic Publishers Group
Post Office Box 322
3300 AH Dordrecht, THE NETHERLANDS
Telephone 31 786 576 000
Fax 31 786 576 254
E-mail: services@wkap.nl

Electronic Services <http://www.wkap.nl>

Library of Congress Cataloging-in-Publication Data

A C.I.P. Catalogue record for this book is available from the Library of Congress.

Hassin, Refael & Haviv, Moshe/ *TO QUEUE OR NOT TO QUEUE: Equilibrium Behavior in Queueing Systems*

ISBN 1-4020-7203-1

To the memory of my parents, Sara and Avraham Haviv

To my mother and late father, Fella and Shimon Hassin

Contents

Preface

The literature on equilibrium behavior of customers and servers in queuing systems is rich. However, there is no comprehensive survey of this field. Moreover, what has been published lacks continuity and leaves many issues uncovered.

One of the main goals of this book is to review the existing literature under one cover. Other goals are to edit the known results in a unified manner, classify them and identify where and how they relate to each other, and fill in some gaps with new results. In some areas we explicitly mention open problems. We hope that this survey will motivate further research and enable researchers to identify important open problems.

The models described in this book have numerous applications. Many examples can be found in the cited papers, but we have chosen not to include applications in the book. Many of the ideas described in this book are special cases of general principles in Economics and Game Theory. We often cite references that contain more general treatment of a subject, but we do not go into the details.

For each topic covered in the book, we have highlighted the results that, in our opinion, are the most important. We also present a brief discussion of related results. The content of each chapter is briefly described below.

Chapter 1 is an introduction. It contains basic definitions, models and solution concepts which will be used frequently throughout the book. This chapter also deals in depth with a seemingly simple model (the shuttle example) which is used to illustrate some of the main themes of this book.

Chapter 2 studies the basic model in which customers decide whether or not to join a queue, after observing its length. The differences between individual optimization (Nash Equilibrium), social optimization, and profit maximization are emphasized. Various ways to regulate the queue and induce customers to behave in the socially desired way are discussed.

Chapter 3 deals with the same model as Chapter 2 except that customers cannot observe the queue length before they make their decisions. We also discuss models with additional features such as: customers know their exact service requirement; the customer population is heterogeneous; the queueing discipline is not first-come first-served.

Chapter 4 analyzes queues in which customers differ by their priority levels. In some models priority is set according to the customer's type, in others customers have the option of buying priority. A main issue in models of the latter type is how to select prices that induce customers to buy the right priority level so that the overall welfare is maximized.

Chapter 5 is concerned with two types of behavior. In the first, customers have the option to abandon (or renege) the queue after their waiting conditions deteriorate. In an observable system customers may renege if the queue becomes too congested. In an unobservable system reneging may result from waiting costs that increase in time. The second type of behavior allows customers to jockey among queues and to purchase information about which queue is the shortest.

Chapter 6 deals with models in which customers possess information on the state of the queue at a given point in time. Examples are service systems that open and close at given times, scheduled service, and models in which customers may leave the system temporarily after observing a long queue and retry at a later time.

Chapter 7 studies competition among service providers who try to attract customers while maximizing their profits. We discuss how prices, priorities, and information are used to achieve this goal.

Chapter 8 deals with long-run decisions of servers regarding their service rates. A higher rate means better service and helps to attract more customers, but it usually comes at a higher cost to the server.

A central issue in this book is how to reduce the time individuals spend waiting in queues. This is a desirable goal since waiting time is often assumed to be fruitless, even though this is not always the case. One of the authors (R.H.) recalls such an incident more than twenty years ago. While waiting in a queue, he first learned about Naor's work from his friend and colleague Ami Glazer from the University of California, Irvine. It was then that the first seed for this book was sown.

Chapter 1

INTRODUCTION

Customers in service systems act independently in order to maximize their welfare. Yet, each customer's optimal behavior is affected by acts taken by the system managers and by the other customers. The result is an aggregate "equilibrium" pattern of behavior which may not be optimal from the point of view of society as a whole. Similar observations have been known to economists for a long time, but have been made explicit in the context of queueing theory only after the publication in 1969 of a paper by P. Naor [133]. The scope of queueing theory prior to Naor's paper is well reflected in a "Letter to the Editor" [103] published in 1964 by W.A. Leeman. It ends with the following excerpt:

> It is a bit surprising that in a capitalistic economy, applied queueing theory limits itself to recommendations of administrative measures for the reduction of queues. One might have expected to observe such an approach in a planned economy but not in an economy in which prices and markets play so large a role.

Leeman saw three objectives that can be attained by pricing a queueing system. First, improving the allocation of existing service facilities by shifting demand from spatial-temporal bottlenecks and allocating through centrally established priorities, rather than according to a first-come first-served rule. Second, decentralizing management decisions, and lastly, guiding long-run investment decisions. Leeman missed a fourth important objective that was filled by the seminal paper of Naor, namely, regulating the demand process that, without such an act, tends to use the facility excessively.

Extensive research on optimal control of queues followed Naor's work. We will concentrate on another area of research that followed Naor's paper, namely, equilibrium behavior in queueing systems. It is interesting to note however, that the concept of equilibrium is not central in Naor's work and is treated there only implicitly (see §2).

A basic economic principle states that the optimal allocation of scarce resources requires that a cost be charged to the users of such resources. Knudsen [90] observed that in stochastic queueing models, the meaning of scarcity is broader than in the usual static and deterministic models of economic theory. If the expected demand for service is smaller than the capacity of the service system, then in a deterministic model the resource is not considered scarce. Still, a cost charged to customers may increase social welfare. This is because *at any point in time there is a positive probability that the service capacity is fully utilized.* Even when the arrival rate is smaller than the service rate (so that the server can accommodate all arrivals), queues are formed due to the variability in service and inter-arrival times. A queue may be considered a price the system has to pay in order to guarantee some level of server utilization. Thus, the criteria for economic optimality are significantly different in stochastic and deterministic models.

In the rest of this introduction we define and discuss the concepts that will be used throughout this book. We also introduce a simple model that illustrates many of the subtleties of decision making in queues.

1. Basic concepts

The concept of an equilibrium plays a central role in this book and the necessary background material is presented in this section.

- Throughout this book "equilibrium" means "Nash equilibrium".

1.1. Strategies, payoffs, and equilibrium

A *non-cooperative game* is defined as follows. Let $N = \{1, \ldots, n\}$ be a finite set of *players* and let A_i denote a set of *actions* available to player $i \in N$. A *pure strategy* for player i is an action from A_i. A *mixed strategy* corresponds to a probability function which prescribes a randomized rule for selecting an action from A_i. Denote by S_i the set of strategies available to player i.

A *strategy profile* $s = (s_1, \ldots, s_n)$ assigns a strategy $s_i \in S_i$ to each player $i \in N$. Each player is associated with a real *payoff function* $F_i(s)$. This function specifies the payoff received by player i given that the strategy profile s is adopted by the players. Denote by s_{-i} a profile for the set of players $N \setminus \{i\}$. The function $F_i(s) = F_i(s_i, s_{-i})$ is assumed to be linear in s_i. This means that if s_i is a mixture with probabilities α and $1-\alpha$ between strategies s_i^1 and s_i^2, then $F_i(s_i, s_{-i}) = \alpha F_i(s_i^1, s_{-i}) + (1-\alpha)F_i(s_i^2, s_{-i})$ for any s_{-i}.

Strategy s_i^1 is said to *weakly dominate* strategy s_i^2 (for player i), if for any s_{-i}, $F_i(s_i^1, s_{-i}) \geq F_i(s_i^2, s_{-i})$ and for at least one s_{-i} the inequality is strict. A strategy s_i is said to be *weakly dominant* if it weakly dominates all other strategies in S_i. A strategy s_i^* is said to be a *best response* for player i against the profile s_{-i} if

$$s_i^* \in \arg\max_{s_i \in S_i} F_i(s_i, s_{-i}).$$

A strategy profile s^e is an *equilibrium* profile if for every $i \in N$, s_i^e is a best response for player i against s_{-i}^e, i.e.,

$$s_i^e \in \arg\max_{s_i \in S_i} F_i(s_i, s_{-i}^e), \quad i \in N.$$

REMARK 1.1 If a best response s_i^* is a mixture of strategies then all these strategies are also best responses. This property does not hold when "best response" is replaced by "equilibrium".

We will deal mostly with games with indistinguishable infinitely many players (usually customers). In this case, denote the common set of strategies and the payoff function by S and F, respectively. Let $F(a, b)$, be the payoff for a player who selects strategy a when everyone else selects strategy b. A *symmetric equilibrium* is a strategy $s^e \in S$ such that

$$s^e \in \arg\max_{s \in S} F(s, s^e).$$

In other words, s^e is a symmetric equilibrium if it is a best response against itself.

We do not assume that an equilibrium always exists. Indeed, in §2.11, §5.1 and §7.3 we present models where no equilibrium exists.

We will often classify queues according to whether or not their length can be observed before a customer makes a decision. We refer to these cases as *observable queues* and *unobservable queues*, respectively. . In observable queues, customers face situations which correspond to *states* of the system and are called upon to choose an action out of a given set. The definitions of actions, strategies, payoffs and equilibria can be extended to state dependent models as well.

For example, a state may correspond to the number of customers in the system, and the action set may include joining as an ordinary customer, joining as a priority customer, or not joining at all. A pure strategy prescribes an action to each state. A strategy profile and an initial state induce a probability distribution over the states. Player i

obtains a payoff that depends on the state, his action, and the strategies selected by others. Player i is interested only in his expected payoff, where the expectation is taken over the states and the actions prescribed by the strategy of customer i in each state.

1.2. Steady-state

When evaluating an individual's expected payoff which is associated with a strategy x as a response against all others using strategy y, we assume that steady-state conditions (based on all using strategy y) have been reached. In most of the models there is an underlying Markov process, whose transition probabilities are induced by the common strategy selected by all. Hence, "steady-state" has the standard meaning of limit probabilities and an individual assumes that this is the distribution over the states.

To illustrate this point assume an $M/M/1$ queue (see Section 4 below) with a potential arrival rate of 4 customers per unit of time, a service rate of 5 per unit of time, and customers who join with probability 0.75. Under steady-state, a customer who considers joining the queue evaluates his expected waiting time by $\frac{1}{5-4\cdot 0.75}$ (see (1.4)). Of course, had this customer been the first or second to arrive to a system that initializes with an empty queue, his evaluation would have been different.

The situation is more involved when the decision maker may face one out of several possible states. For example, consider the observable version of the above decision problem in which customers observe the queue length before deciding whether or not to join. Consider a strategy δ and denote the action taken according to it in state s by $\delta(s)$. For simplicity, assume that δ is a deterministic strategy. Here $s = 0, 1, 2, \ldots$ are possible queue lengths an arrival may face upon arriving, and $\delta(s)$ can either be *join* or *balk*. Such strategy, when used by all, determines the transition probabilities over a Markov process with state space $\{0, 1, 2, \ldots\}$. Let $\pi_s(\delta)$ be the limit probability of state s given that s is the initial state and strategy δ is adopted by all.[1] Hence, the expected waiting time for an individual who uses strategy δ' when all use strategy δ is

$$\sum_{s|\delta'(s)=join} \pi_s(\delta)\frac{s+1}{\mu}. \tag{1.1}$$

[1] In case of periodicity, with period d, replace the limit by averaging the limits along d consecutive periods. Note that $\sum_{s=0}^{\infty} \pi_s(\delta)$ does not necessarily sum up to 1. On one hand, it can be greater than 1 (in fact, can even be unbounded) when more than one recurrent chain exists, and on the other hand it may sum up to 0. An example for the latter case is when $\lambda > \mu$ and $\delta(s) = join$ for all $s \geq 0$.

1.3. Subgame perfect equilibrium

A commonly cited drawback of the equilibrium concept is the possibility that the solution is not unique. We describe here and in the next subsection two refinements which can be used to reduce the number of solutions.

The transition probabilities between various states usually depend on the strategy adopted by the customers. In particular, it is possible that for a given strategy and initial state, some states have zero steady-state probability. When computing the customers' expected payoffs, these states receive a weight of 0. Therefore, it is immaterial which actions are prescribed for these states in order to examine whether a given strategy is a best response. For example, for those states s with $\pi_s(\delta) = 0$, the value of (1.1) is the same regardless of whether $\delta'(s)$ is *join* or *balk*. Yet, a strategy ought to prescribe an action for every state. This fact often leads to multiple equilibria, some of which are counterintuitive.

A *subgame perfect equilibrium* (SPE) prescribes best responses in all states, including those that have zero steady-state probability. An example for multiple equilibria with exactly one SPE is given in Section 5.2. For more on the concept of SPE in queueing systems see Hassin and Haviv [73].

1.4. Evolutionarily stable strategies

A (symmetric) equilibrium strategy is, by definition, a best response against itself. However, it need not be the unique best response. Specifically, let y be an equilibrium strategy. There may be a best response strategy $z \neq y$ such that z is strictly a better response against itself than y is. In this case, y is unstable in the sense that when starting with y, it may be that the players adopt the best response z, and then a new equilibrium, at z, will be reached. If no such z exists then y is said to be an *evolutionarily stable strategy* or ESS (see Maynard-Smith [122]). Note that if y is an equilibrium strategy and it is the unique best response against itself, then it is necessarily ESS.

Formally, an equilibrium strategy y is said to be an ESS if for any $z \neq y$ which is a best response against y, y is better than z as a response to z itself: $y \in \arg\max_{x \in S} F(x, y)$, and for any strategy $z \neq y$ such that $z \in \arg\max_{x \in S} F(x, y)$, $F(y, z) > F(z, z)$. Note, that there exist examples in which no equilibrium strategy is an ESS.

1.5. The Braess paradox

The addition of new options may lead to a new equilibrium in which everybody is worse-off. The following payoff matrix describes an instance of the well known *prisoner's dilemma* (where (x, y) means a payoff of x to the row player and y to the column player):

	A	B
A	(1,1)	(3,0)
B	(0,3)	(2,2)

In the unique equilibrium, both players select A. Yet, if they select B instead, both get higher payoffs. If B were the only option, they would both end up with 2, but once option A is introduced, the resulting new equilibrium is worse for both.

Braess [30] introduced this phenomenon in the context of transportation models, showing that the addition of a new road segment may lead to an equilibrium in which all users of a road network are worse-off. This phenomenon is commonly denoted as the *Braess paradox.*

The paradox may also appear when an increase in the amount of information available to the players leads to a new equilibrium in which all are worse-off. Indeed, more information may mean more strategies, so that this phenomenon is in line with the Braess paradox. The effect of increased information and the Braess paradox in queueing systems, are discussed in §3.2 and §3.8.

1.6. Avoid the crowd or follow it?

In many queueing models, strategies can be represented by a single numerical value. For example, in the "bribery" model of §4.5, a strategy prescribes how much to pay for service. In such cases, the following question turns out to be meaningful:

> **Is an individual's best response a monotone increasing (or decreasing) function of the strategy selected by the other customers?**

Let $F(x, y)$ be the payoff for a customer who selects strategy x when all others select strategy y. Assume that for any y there exists a unique best response $x(y)$:

$$x(y) = \arg\max_x F(x, y).$$

We are interested in cases where $x(y)$ is continuous and strictly monotone. Figure 1.1 illustrates a situation where a strategy corresponds to a nonnegative number. It depicts one instance where $x(y)$ is monotone increasing and another where it is monotone decreasing. We call these situations *avoid the crowd* (ATC) and *follow the crowd* (FTC), respectively. The rationale behind this terminology is that in an FTC (respectively, ATC) case, the higher the values selected by the others, the higher (respectively, lower) is one's best response.

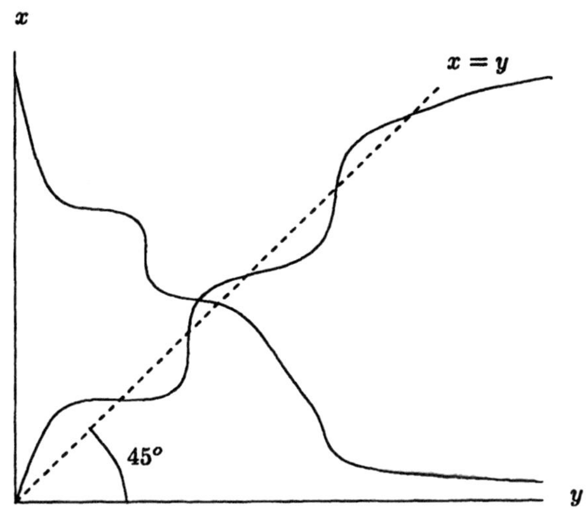

Figure 1.1. ATC and FTC instances

An equilibrium strategy y satisfies $x(y) = y$. In other words, it is a fixed point of the function x. It is of interest to determine if a model is ATC or FTC since, clearly, in the ATC case at most one equilibrium exists whereas multiple equilibria are possible in the FTC situation.

2. Threshold strategies

In this section we describe a class of strategies, known as *threshold strategies*, which is common in queueing systems. Suppose that upon arrival the customer has to choose between two actions, A_1 and A_2, after observing a nonnegative integer-valued variable which characterizes the state of the system. For example, the state may be the length of the queue and the actions may be to join or to balk.

A *pure threshold strategy* with threshold n prescribes one of the actions, say A_1, for every state in $\{0, 1, \ldots, n-1\}$ and the other action, A_2, otherwise.

In many cases it is natural to look for an equilibrium pure threshold strategy. However, it is often possible to construct instances where, for example, if everyone in the population uses the threshold 4 then the best response for an individual is 5 and if everyone in the population adopts the threshold 5 then the best response is 4. This is the case with the upper function in Figure 1.2. In such cases, a pure threshold strategy that defines an equilibrium may not exist. Consequently, the definition of a threshold strategy is extended as follows:

A threshold strategy with threshold $x = n + p$, $n \in I\!N$, $p \in [0, 1)$, prescribes mixing between the two pure threshold strategies n and $n+1$ so that strategy n receives the weight of $1-p$ and strategy $n+1$ receives the weight of p. The resulting behavior is that all select a given action, say A_1, when the state is $0 \leq i \leq n-1$; select randomly between A_1 and A_2 when $i = n$, assigning probability p to A_1 (the action prescribed by strategy $n+1$) and probability $1-p$ to A_2 (the action prescribed by strategy n); select A_2 when $i > n$. If x is an integer ($p = 0$), the strategy is *pure*. Otherwise, it is *mixed*.

We are interested in models where a best response for an individual against any strategy x is a pure threshold strategy: for some integer $k(x)$, if the state is in $\{0, \ldots, k(x) - 1\}$ choose A_1. Otherwise, choose A_2. The following situation is typical: $k(x)$ has points of discontinuity with a step of unit size which may be upwards or downwards. At a point of discontinuity x, both of the two pure strategies involved are best responses against x and hence any mixing between them (which is a mixed threshold strategy) is also a best response against x. A threshold x defines an equilibrium if either $k(x) = x$ (in which case x is an integer) or x is between $k(x-)$ and $k(x+)$.[2] In both cases, if all customers adopt the threshold strategy x, then this is also a best response and no one has an incentive to deviate to another strategy. In short, it is an equilibrium strategy.

Recalling from Section 1.6, the behavior reflected by a monotone non-increasing function $k(x)$ is referred to as *avoid the crowd* (ATC). It means that the higher is the threshold adopted by others, the lower is the threshold giving the best response for a given customer. Similarly, the case where $k(x)$ is monotone non-decreasing is referred to as *follow the crowd* (FTC). It means that the higher the threshold adopted by others, the higher is the threshold giving the best response for a given customer.

There are important differences between the two cases. Under ATC there is at most one fixed point. It may describe a pure strategy or a mixed one. Figure 1.2 depicts two non-increasing step functions. In one, the equilibrium strategy obtained at x_1 is pure, in the other, the equilibrium strategy obtained at x_2, is mixed. The FTC case is more involved and it may have multiple equilibria. It can be seen from Figure 1.3 that $k(x)$ may have numerous fixed points.

REMARK 1.2 The data are said to be *non-degenerate* if none of the jumps of $k(x)$ occur at an integer x value. Let $x_1, x_2, \ldots$ be the values of

[2] It is convenient to view both cases as solutions to the equation $k(x) = x$, that is, as fixed points of $k(x)$.

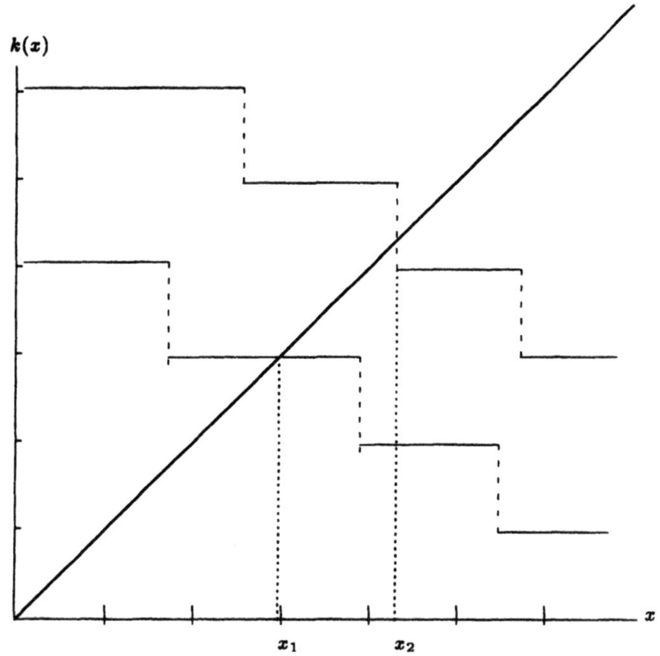

Figure 1.2. Equilibrium in an ATC situation

the fixed points. From Figure 1.3 we observe that for $k = 1, 2, \ldots$, x_{2k+1} corresponds to a pure equilibrium strategy whereas x_{2k} corresponds to a mixed equilibrium strategy. When we allow degenerate data there may be consecutive pure equilibrium strategies. If the equilibrium is unique then it is pure.

3. Costs and objectives

The welfare of a customer consists of benefits associated with service, from which direct payments and indirect costs associated with waiting are subtracted. The sum of direct and indirect costs is referred to as the *full price.* We assume that the customers involved are risk neutral in the relevant range of payments and benefits so that they maximize their expected welfare.

In most cases it is assumed that the value of a unit of time for each customer is constant (denoted by C), so that spending t time units in

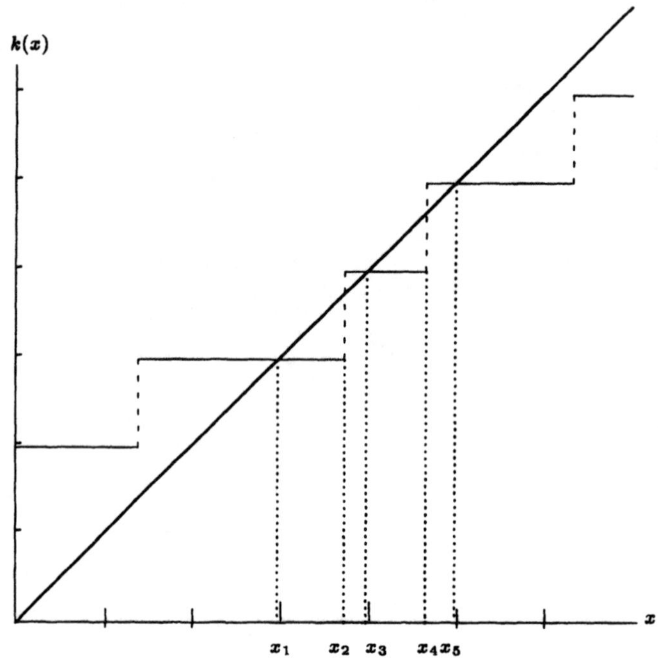

Figure 1.3. Equilibrium in an FTC situation

the system has a total cost of Ct.[3] The value of C may differ from one customer to another.[4]

A queueing system may also be considered from a social point of view. When we adopt this viewpoint, we assume that the goal in controlling the system is to maximize *social welfare* which is defined here as the total expected net benefit of the members of the society, including both customers and servers. From this approach, a payment transferred between individuals in the population has a zero net effect on social welfare and thus no effect on the system's optimization. Therefore, the social goal is to maximize the sum of benefits from service minus waiting and operating costs.

[3]An interesting generalization to this rule is proposed by Balachandran and Radhakrishnan [19]. Suppose that waiting t time units costs Ce^{at} for given parameters $C > 0$ and $a \geq 0$. Then, the expected waiting cost of a customer is $\int_0^\infty Ce^{at}w(t)\,dt$ where $w(t)$ is the density function of the waiting time. In an $M/M/1$ system $w(t) = (\mu - \lambda)e^{-(\mu-\lambda)t}$ where λ is the arrival rate and μ is the service rate. In this case the expected cost equals $\frac{C}{\mu-a-\lambda}$. Note that the case of linear waiting costs is obtained when $a = 0$.

[4]See Deacon and Sonstelie [43] and Png and Reitman [140] for empirical studies concerning this parameter.

In some cases we will deal with models of *class decision*, where customers belong to classes and each class makes its decisions to maximize the total welfare of its members. This assumption leads to a noncooperative game with a finite number of players. It is assumed that the arrival processes of customers in various classes are independent, and in particular, if the joint arrival process is Poisson with rate λ, and the proportion of class-i customers is p_i, then the arrival process of i-customers is Poisson with rate $p_i\lambda$.

We use the term *waiting time* for the time from arrival to departure. Some authors use the term *sojourn time*. Waiting time excluding service time is referred to as *queueing time*.

4. Queueing theory preliminaries

This section contains a short account of some basic concepts and results from queueing theory which will be used frequently in this book. We use conventional notation to describe basic models. For example, an $M/G/s$ system has s identical servers facing a Poisson stream of customers and no specific service distribution is assumed (M stands for "Markovian" whereas G stands for "general"). The quoted results assume steady-state conditions.

We consider a variety of types of decisions made by the customers of a queueing system. A main one is whether to join or not. We apply the common terminology that distinguishes between *balking* as the act of refusing to join a queue and *reneging* as the act of leaving a queue after joining it. The *arrival process* usually refers to the process by which the demand for service is generated, whereas the *joining process* consists only of those customers who decide to join (i.e., they do not balk). In the literature, the rates of arrival and joining are often termed as the *potential demand*, and the *effective arrival rate*. When the arrival and joining rates differ, that is when balking is exercised, we often use Λ for the arrival rate and λ for the joining rate.

The service discipline mostly discussed in this book is *first-come first-served* (FCFS). However, we frequently deal with other regimes. There are two common versions of *last-come first-served* (LCFS) disciplines. The first, *without preemption*, in which a new arrival is positioned at the head of the queue but the customer in service is allowed to complete it. The second, *with preemption*, in which a new arrival preempts a customer who might be in service. It is usually assumed that service, when resumed, is continued from the point where it was interrupted. The acronym used to describe this discipline is LCFS-PR.

Another queueing regime is *processor sharing*. It has two common versions. Under *egalitarian processor sharing* (EPS) the server splits its service capacity evenly among all present customers. In particular, if n customers are present during the entire time interval of length Δt, and at the beginning of this interval their completed workloads were $(x_1, \ldots, x_n)$, then at its end their completed workloads are $(x_1 + \frac{\Delta t}{n}, \ldots, x_n + \frac{\Delta t}{n})$. Also, if the service requirements are exponential with rate μ, then a tagged customer completes his service during the next Δt units of time with probability $\frac{\mu}{n}\Delta t + o(\Delta t)$. Otherwise, when the split of capacity is based on customers' parameters, it is called *discriminatory processor sharing* (DPS). There are also two versions of *random order* disciplines. In one, whenever a server becomes free, a random customer from the queue is selected to commence service. In the other version, whenever a server completes service, a customer from the queue is randomly chosen to be the one to whom this service is granted. We will mention some similarities between EPS and the latter type of the random order discipline.

A service discipline is *strong* if the rule by which the next customer to be served is selected does not take into account the actual residual service requirements. It is *work-conserving* if the server is never idle when the queue is not empty, and a customer whose service was interrupted resumes it from the point of interruption. Under a work-conserving discipline, the total unfinished work at any time is the same as in the corresponding FCFS model.

Examples for disciplines that are strong and work-conserving are FCFS, LCFS, random order, order which is based on customers payments, and EPS.

Service requirements are assumed to be independent and identically distributed. Denote by μ^{-1} the (common) expected service requirement (i.e., μ is the *rate of service*). For stability, assume that the system's *utilization factor* $\rho = \frac{\lambda}{\mu}$ is strictly less than 1 (sometimes, when individual optimization leads to stability, this assumption is removed).

The following five results hold when the arrival process is Poisson with rate λ, the service distribution is exponential (an $M/M/1$ model) with rate μ, and the service discipline is strong and work-conserving. They also hold for $M/G/1$ models when the service discipline is either EPS or LCFS-PR.

- The probability that n ($n \geq 0$) customers are in the system (at arbitrary times as well as at arrival times) is

$$(1 - \rho)\rho^n. \tag{1.2}$$

- The expected number of customers in the system is

$$\frac{\rho}{1-\rho}. \tag{1.3}$$

- The expected waiting time is

$$\frac{1}{\mu(1-\rho)} = \frac{1}{\mu-\lambda}. \tag{1.4}$$

- The expected length of a *busy period*, i.e., the expected time from an arrival to an idle server until the server becomes idle again, is[5]

$$\frac{1}{\mu(1-\rho)}. \tag{1.5}$$

- The expected time between a customer's arrival and the first time the server is idle is

$$\frac{1}{\mu(1-\rho)^2}. \tag{1.6}$$

The following property holds for $M/M/1$ queues:

- The time it takes to reduce the number of customers in the system from n to $n-1$, $n \geq 1$, and the length of a busy period are identically distributed.

For a general service distribution we obtain the $M/G/1$ model. The Khintchine-Pollaczek (K-P) formula calculates the expected queueing time in an $M/G/1$ queue where the service discipline is strong, work-conserving and without preemption. Examples for such disciplines are FCFS, LCFS without preemption, and random order. Examples for disciplines for which the K-P formula does not hold are EPS and LCFS-PR. The K-P formula says that the expected queueing time (excluding service) is

$$W_q = \frac{\lambda\overline{x^2}}{2(1-\rho)}, \tag{1.7}$$

where $\overline{x^2}$ denotes the mean squared service time.

[5]This expression is identical to (1.4). It can be explained by observing that in a LCFS-PR $M/M/1$ system, an arrival stays for a length of time that has the same distribution as a busy period. Moreover, all strong and work-conserving disciplines in $M/M/1$ queues share the same expected waiting time.

5. A shuttle example

In this section we illustrate some of the issues presented in the previous section.[6] The model is concerned with two servers operating according to different modes of batch service. For convenience, we present the model in terms of transportation services. Consider two types of such services. The first is a shuttle service that departs whenever the number of waiting passengers reaches the transporter's capacity.[7] We assume the capacity of a transporter is seven passengers. No limit on the number of transporters is assumed. The second is a bus service. Buses arrive according to a stochastic process with a known expected *residual* inter-arrival time[8] which we assume to be five minutes. No limit on the bus capacity is assumed.

Commuters are assumed to have no preference as to the type of service, and their only concern is their expected waiting times. The generation of commuters who require service is assumed to form a Poisson process with rate λ. We analyze the commuters' decision process under two different cases. In the observable model, an arriving commuter observes the number of waiting commuters at the shuttle service, whereas in the unobservable model he does not have this information.

5.1. The unobservable model

Suppose that when making their choice, commuters are not informed about the number of commuters at the shuttle terminal at that instant. Also, once a decision is made, it is too costly to change it.

The waiting time of a commuter who selects the shuttle service depends on the choice made by the others. Specifically, the higher the rate of arrival to the shuttle station, the lower the expected waiting time for this service. Thus, if a critical mass of the commuters chooses the shuttle service, then the expected waiting time until the shuttle departs might be sufficiently low, making it attractive for individual commuters. This, of course, is possible only if λ is not too small. We will show that the precise condition is $\lambda > \frac{3}{5}$, which is assumed below. Consequently, we expect one of the following situations: either all choose the shuttle service, or none do.

---

[6]The model which we describe in this section is an example of a *coordination game*. See, for example, Crawford [38].

[7]Compare with Kosten's "unscheduled ferry problem" and "custodian's problem" [93].

[8]The residual inter-arrival time is the period between a random inspection of the process until an arrival occurs.

Both solutions define equilibria and nothing in the description of the model can determine which solution will be obtained. Moreover, as we show next, both are ESS.

A strategy, pure or mixed, corresponds to a fraction, p, which is the probability of selecting the shuttle service (so that $1-p$ is the probability of selecting the bus operation). Let $W(p,q)$ be the expected waiting time for an individual who uses strategy p, while the others use strategy q. Strategy p is an ESS if the following two conditions hold:

- $W(p,p) \leq W(q,p)$ for any strategy q;
- if $W(q,p) = W(p,p)$ for $q \neq p$, then $W(p,q) < W(q,q)$.

Both $p = 0$ and $p = 1$ are ESS, since $W(1,1) < W(q,1)$ (respectively, $W(0,0) < W(q,0)$) for any $q \neq 1$ (respectively, $q \neq 0$) and hence the second condition for an ESS is automatically satisfied.[9]

Next we examine whether additional equilibria, obviously mixed, exist. Suppose that commuters use a mixed (symmetric) strategy with $0 < p < 1$. The resulting arrival process to the shuttle is Poisson with rate $p\lambda$. This strategy defines an equilibrium if commuters are indifferent between using the shuttle and the bus, so that no commuter has an incentive to change his behavior. However, such an equilibrium is not an ESS. Indeed, if any fraction (let alone all) of the commuters changes its behavior, say by making the bus service more popular, this will tip the balance and make the bus service more attractive. In other words, deviation from the equilibrium is self-perpetuating, and it causes more (and more) to deviate *in the same direction*, leading to a new equilibrium in which all use the bus. Similarly, any shift to a more frequent use of the shuttle service will result in everybody using this service.

Recall that the shuttle capacity is 7 and the expected waiting time for a bus is 5 minutes. Assume that the commuters follow the mixed strategy p. If a tagged commuter selects the shuttle service, then the number of waiting commuters that he meets upon arrival is i with probability $\frac{1}{7}$ for $i = 0, \ldots, 6$, and his expected waiting time equals $\frac{6-i}{\lambda p}$.[10] Therefore, his (unconditional) expected waiting time is $\frac{3}{\lambda p}$. If $p < \frac{3}{5\lambda}$ (respectively, $p > \frac{3}{5\lambda}$) then his unique best selection is the bus (respectively, shuttle).

[9] If, for example, the shuttle operator prefers the solution with $p = 1$, then one way to attain this goal is by convincing the commuters that this is indeed the situation!

[10] This is where the steady-state assumption is used. The number of commuters waiting for the shuttle service defines a Markov process with the state space $S = \{0, 1, \ldots, 6\}$. The only transitions are from state i to state $(i+1)$ *mod* 6 and they all have rates λp. The resulting limit probabilities are uniform.

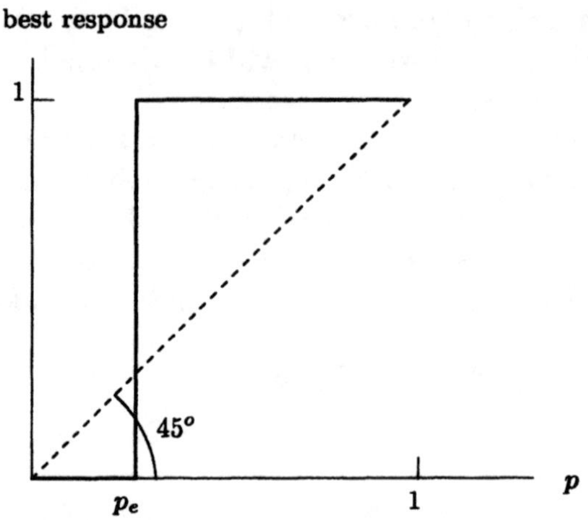

Figure 1.4. Best response vs. fraction of shuttle users

He is indifferent between the two options if $p = p_e$ where $p_e = \frac{3}{5\lambda}$.[11] Thus, p_e is an equilibrium strategy. However, p_e is not an ESS. For example, $W(0, p_e) = W(p_e, p_e)$ but $W(0,0) < W(p_e, 0) = \infty$. A similar result holds when 0 is replaced with 1. In fact, p_e is a best response for an individual only if all others use this strategy, whereas any strategy p is a best response against p_e. Note that if all use $p > p_e$ then 1 is uniquely the best response, and if $p < p_e$ then 0 is uniquely the best response (see Figure 1.4).

REMARK 1.3 The best response function is monotone non-decreasing. This is an example of an FTC situation, which explains the existence of multiple equilibria.

To summarize, three equilibrium strategies exist in the non-trivial case where $\lambda > \frac{3}{5}$. The two pure equilibria are ESS. Moreover, $p = 0$ remains the unique best response as long as the rest use strategy $p < \frac{3}{5\lambda}$. Similarly, $p = 1$ is the unique best response as long as the the rest use strategy $p > \frac{3}{5\lambda}$. The stability of the equilibria $p = 0$ and $p = 1$ is reflected by the property that even if a non-negligible deviation from strategies 0 or 1 takes place among all commuters, the best response is not affected. The mixed equilibrium prescribes the shuttle with probability $p_e = \frac{3}{5}$. This equilibrium is not an ESS.

[11] When $\frac{3}{5\lambda} > 1$, commuters appear at a rate so low that even when all of them use the shuttle service, the individual's best response is still to use the bus service. In other words, when $\lambda < \frac{3}{5}$, using the bus service is a dominant strategy.

5.2. The observable model

Assume now that commuters observe the number of people already waiting at the shuttle terminal, and then decide which service to choose. Of course, the higher the number of commuters already waiting for the shuttle, the more a new arrival tends to select this service.

Consider an individual who arrives when the shuttle is empty, and assume the most favorable case in which all select this service. His expected waiting time for the shuttle service is then $\frac{6}{\lambda}$. If under these conditions it is best for him to select the shuttle service, it is of course also best for those who observe a longer queue for the shuttle service. On the other hand, if he does not select the shuttle service, nobody will (the next arrivals will also face an empty queue and will follow through with the same reasoning), and hence this service will never be used. The conclusion of this analysis is that if $\lambda > \frac{6}{5}$, all use the shuttle service and if $\lambda < \frac{6}{5}$, none do.

This informal analysis is correct but not complete. In particular, it does not distinguish between equilibria which are subgame perfect and those which are not. We now proceed with a formal analysis.

A pure strategy is a function from the number, i, of commuters waiting for the shuttle upon arrival of a commuter to the set of actions (i.e., which service to select). For example, the strategy $\delta = (s, b, s, b, s, b, s)$ prescribes taking the shuttle whenever i is even and taking the bus whenever i is odd. It is clear that those states where b is prescribed are recurrent (in fact, absorbing) and those where s is prescribed are transient. The exception for the latter is $\delta = (s, s, \ldots, s)$ where all states are recurrent.[12]

When $\lambda > \frac{6}{5}$, $(s, s, \ldots, s)$ is the unique equilibrium. Moreover, as all states are recurrent this is also the unique SPE. The analysis is more involved when $\lambda < \frac{6}{5}$. Now the strategy $(s, s, \ldots, s)$ is not an equilibrium: given that all follow it, it prescribes a suboptimal action for the recurrent state $i = 0$. Therefore, a necessary condition for a pure strategy δ to be an equilibrium is that $i(\delta) \geq 0$ where

$$i(\delta) = \max\{i | \delta(i) = b\}.$$

As said above, all states for which the bus is prescribed under δ are recurrent (in fact, absorbing), whereas the others are transient. Thus, for

[12]A state is said to be ***recurrent*** if a Markov process which initiates in it will visit it again with probability 1. Alternatively, this state will be visited an infinite number of times with probability 1. A recurrent state is said to be ***absorbing*** if once a process enters it, it will stay there forever with probability 1. A non-recurrent state is said to be ***transient***. In other words, a Markov process which initiates in it will visit it again with probability less than 1, and thus the number of visits there is finite with probability 1.

an equilibrium, only optimality at the former group has to be checked.[13] In fact, only optimality at $i(\delta)$ has to be checked since if b is an optimal action at $i(\delta)$, it is certainly an optimal action for states i such that $i < i(\delta)$, and if $\delta(i) = s$ for $i \neq i(\delta)$ then i is transient. In particular, δ is an equilibrium if and only if $\frac{6-i(\delta)}{\lambda} \geq 5$. In other words, the set of pure equilibria are all δ with $0 \leq i(\delta) \leq 6 - 5\lambda$.

Clearly, in a SPE, s is prescribed for state 6. Given that $i > 6 - 5\lambda$ and that for every state j, $j > i$, the strategy prescribes s, then to be a best response, the strategy must also prescribe s to state i. Likewise, for states $i \leq 6 - 5\lambda$ it must prescribe b. Thus, an equilibrium strategy is subgame prefect if and only if it prescribes the bus for all states $i \leq 6 - 5\lambda$ and the shuttle for the rest.

To summarize, excluding the cases where $6 - 5\lambda$ is an integer, for any λ there exists a unique SPE which is of the type $(b, \ldots, b, s, \ldots, s)$.[14] When $6 - 5\lambda \geq 2$, more equilibria exist and they are of the form $(x, \ldots, x, b, s, \ldots, s)$ where x stands for any action and the last b appears in some position i where $i \leq 6 - 5\lambda$.

We conclude the equilibrium analysis by comparing the unobservable and the observable cases. When $\frac{3}{5} < \lambda < \frac{6}{5}$, the shuttle operator is better-off in the unobservable case than in the observable case. Indeed, in the unobservable case, there is one ESS in which the shuttle operator gets all of the demand whereas in the observable case, the shuttle operator gets no demand at all in all possible equilibria. If the option is given, the shuttle operator would conceal the information on how many commuters are waiting. The opposite holds when $\lambda > \frac{6}{5}$. Here, when the information is concealed there exists an ESS where the shuttle operator receives no demand. Therefore, the shuttle operator prefers to reveal the information and get all the demand in the unique equilibrium (in fact, SPE).[15] When $\lambda < \frac{3}{5}$, the shuttle operator is indifferent between the options since in both cases all of the demand goes to the bus operator.

5.3. Social optimality

We now illustrate the difference between the equilibrium solution and the solution that maximizes the overall welfare of the commuters. In our model, a commuter who chooses the shuttle generates *positive externalities*. This means that by increasing the rate of arrival to the shuttle, the expected waiting time of the other commuters who make the same

[13]The steady-state assumption is used here: under these conditions, the transient states have zero probability of being observed by an arrival.

[14]The set of b's may be empty, when $\lambda > \frac{6}{5}$, but not the set of s's.

[15]Compare with §3.2.

choice is reduced (whereas the expected waiting time of the bus users is not affected). Customers who independently maximize their own welfare ignore these externalities, and it therefore may happen that the shuttle is less used in equilibrium than under a social welfare maximizing solution.

The socially optimal solution of the unobservable model is simple. Since the social goal is to minimize the expected waiting time, the solution is either that all commuters use the shuttle or that they all use the bus. In the first case the expected waiting time is $\frac{3}{\lambda}$ and in the second case it is 5. Therefore, if $\lambda > \frac{3}{5}$ then all should use the shuttle, and if $\lambda < \frac{3}{5}$ then all should use the bus. If $\lambda = \frac{3}{5}$ the two solutions are socially equal, meaning that it does not matter which means of transportation the commuters use, as long as they all use the same one. Note that when $\lambda < \frac{3}{5}$ the unique equilibrium solution is socially optimal. When $\lambda > \frac{3}{5}$, social optimality requires that all use the shuttle, but the solution that all use the bus is also an equilibrium.

The social considerations in the observable case are similar to those in the unobservable case, and the same behavior is optimal: if $\lambda > \frac{3}{5}$ then all should use the shuttle, and if $\lambda < \frac{3}{5}$ then all should use the bus. Note that when $\lambda > \frac{6}{5}$, the socially optimal solution where all use the shuttle is also an equilibrium, but in the range $\frac{3}{5} < \lambda < \frac{6}{5}$ the socially optimal solution where all use the shuttle is not an equilibrium. The reason, as we have mentioned before, is that the commuters ignore the positive externalities associated with the action of choosing the shuttle. Among those who ignore the externalities are those who arrive to an empty shuttle: optimizing their individual welfare they prefer the bus but socially it is desired that they join the shuttle's queue.

6. Non-stochastic models

In this section we briefly mention some papers that deal with equilibria in non-stochastic queueing models. When a good is available in limited quantity and sells below market price, queues form. The length of waiting time stabilizes at such a level that the full price (consisting of the good's nominal price plus the waiting cost) of the marginal potential consumer equals the good's value. Thus, the resulting equilibrium waiting time may be independent of the service time. Barzel [26] explained that

> the resource cost of this allocation, time spent in the queue, represents a cost of establishing property rights in the good.

He also concluded that although one expects the poor, who have low time value, to profit from the institution of "rationing by waiting", the beneficiaries may often be among the rich.

Donaldson and Eaton [46] considered price discrimination between consumers with low and high time values by offering to sell the product under two options involving different combinations of money and time prices. We will describe related queueing models in §7.

Another model in Shy [156], is based on MacKie-Mason and Varian [115]. There, the utility of customer i, $i = 1, \ldots, n$, is assumed to be given by

$$U_i = \sqrt{q_i} - C\frac{Q}{U} - pq_i,$$

where q_i is the capacity allocated to (or used by) customer i, $Q = \sum_{j=1}^{n} q_j$, U is the total capacity available, C is a cost parameter, and p is a price charged per unit of capacity. The term $C\frac{Q}{U}$ reflects the effect of congestion. Shy derived the equilibrium as well as the social optimal capacity allocations, and showed that the two coincide when $p = C\frac{n-1}{U}$.

Several other papers in the economic literature deal with rationing of goods through queueing, see [6, 135, 141, 160] and their references. These papers model centrally planned economies in which products can be purchased at a low official price after waiting in queues. The products can also be purchased in "black markets" at higher prices and without waiting. This situation opens possibilities for poor customers to spend their time in queues, buy products at the official prices, and re-sell them in the black market. Thus, each individual allocates his resources, money and time, in order to maximize his utility. This literature treats waiting times in queues in the same way it treats the prices, namely, it searches for prices and waiting times which determine equilibria, and apart from this no queueing mechanism that relates waiting time to supply and demand is assumed. Therefore, such models do not fit the framework of this book.

Chapter 2

OBSERVABLE QUEUES

This chapter deals with queueing systems, where an arriving customer observes the length of the queue before making his decisions.

1. Naor's model

The subject of Naor's paper [133] is the control of a FCFS $M/M/1$ system. In Naor's model, a queue manager announces an admission fee, and customers react by setting a pure strategy which distinguishes the states of the queue where customers join from those where they balk. It is easily seen that individual optimization generally determines an equilibrium based on a pure threshold strategy. The concept of equilibrium is not central in this model since the decision of whether to join the queue in a given state is independent of the strategy adopted by the other customers. Yet, a customer's decision to balk when observing a queue length greater than the threshold, is based on the assumption that those present in the queue will not leave (renege) before they are served (see Remark 2.1 below).

Naor noticed that in observable queues the individual's decision deviates from the socially preferred one. This gap is caused by externalities generated when joining the queue: a customer who joins the queue may cause future customers to spend more time in the system. The individual's objective does not take these externalities into consideration. Because of these negative external effects, the equilibrium arrival rate is greater than the socially desired one.[1] We start by listing the assumptions underlying Naor's model:

1 A stationary Poisson stream of customers - with parameter λ - arrives to a single server facility.

2 The service times are independent, identically, and exponentially distributed with parameter μ.

3 A customer's benefit from completed service is R.

4 The cost to a customer for staying in the system (either while waiting or while being served) is C per unit of time.

5 Customers are risk neutral, that is, they maximize the expected value of their net benefit.

6 Utility functions of individual customers are identical and additive, from the public (social) point of view.

7 $R\mu \geq C$.[2]

8 The service discipline is FCFS.

9 A decision to join is irrevocable, and reneging is not allowed.

10 Upon arrival, a customer inspects the queue length and decides whether to join or balk. A customer who balks leaves the system and never returns.

The individual's optimizing strategy in this model is straightforward. A customer who joins the queue when i customers are already in the system (including the one who is currently served) expects a benefit $R - \frac{(i+1)C}{\mu}$. The customer then enters if this value is nonnegative, that is, if $i + 1 \leq \frac{R\mu}{C}$. Otherwise, the customer balks. Consequently, the pure threshold strategy n_e with[3]

$$n_e = \left\lfloor \frac{R\mu}{C} \right\rfloor, \tag{2.1}$$

is an equilibrium strategy. Under this strategy, an arriving customer joins the queue if he observes $n_e - 1$ or fewer customers and balks if he observes n_e customers or more. Note that n_e is the maximum possible number of customers in the system under individual optimization, and the result is an $M/M/1/n_e$ queueing system.

Overall (social) optimization is not as trivial. We observe that there exists a pure threshold socially optimal strategy. This can be argued as

[1] Individual optimization causes more congestion than is socially desired also in more general models. See, for example, Mills [129].

[2] Otherwise, all individuals, even in the most ideal situation of observing an empty system, would balk.

[3] $\lfloor x \rfloor$ denotes the largest integer which is less than or equal to x.

follows: clearly, a pure optimal strategy exists, and any pure strategy is in effect a threshold strategy, where the threshold coincides with the smallest queue size for which the strategy prescribes balking (see Remark 2.2).

Denote by S_O the expected social benefit per unit of time.[4] Given a maximum queue length of n, the probability of observing n customers in the system is $q_n = \frac{\rho^n}{\sum_{i=0}^{n} \rho^i}$. Assuming $\rho \neq 1$, the probability that an arriving customer joins is[5]

$$1 - q_n = \frac{1-\rho^n}{1-\rho^{n+1}}, \tag{2.2}$$

and the expected number of customers in the system is

$$L_n = \frac{\rho}{1-\rho} - \frac{(n+1)\rho^{n+1}}{1-\rho^{n+1}}.$$

Hence,

$$\begin{aligned} S_O &= \lambda R(1-q_n) - CL_n \\ &= \lambda R \frac{1-\rho^n}{1-\rho^{n+1}} - C\left[\frac{\rho}{1-\rho} - \frac{(n+1)\rho^{n+1}}{1-\rho^{n+1}}\right]. \end{aligned} \tag{2.3}$$

Let n^* be a maximizer of (2.3). Naor designed a procedure for computing n^* which is based on the property that the function given in (2.3) is unimodal.[6] Naor also showed that $n^* \le n_e$. In the next section we present an alternative derivation for n^*.

In order to motivate customers to adopt the threshold n^* rather than n_e, Naor suggested imposing an appropriate admission fee. Based on (2.1), an admission fee p induces the socially optimal threshold if

$$n^* = \left\lfloor \frac{(R-p)\mu}{C} \right\rfloor. \tag{2.4}$$

Payments are not considered part of the social welfare function and therefore the exact fee is irrelevant as long as it satisfies (2.4).

Alternatively, the queue can be regulated by imposing a toll on waiting, i.e., increasing C instead of reducing R. Such a toll, t, induces the

---

[4]The subscript O stands for *observable*.

[5]The results hold also for $\rho = 1$ when taking the appropriate limits. In fact, in this case $1 - q_n = \frac{n}{n+1}$.

[6]Haviv and Puterman [77] showed in a different way that a threshold optimal strategy exists.

optimal threshold n^* if

$$n^* = \left\lfloor \frac{R\mu}{C+t} \right\rfloor.$$

We will derive n^* in Section 3.

REMARK 2.1 If Assumption 9 is relaxed and customers are allowed to renege, the decision of whether or not to join may depend on the behavior of the customers who are already in the system. If some of them plan to renege, then a customer may join even if he observes n_e or more customers in the queue. Yet, for a customer who joined after observing at most $n_e - 1$ customers, reneging later is clearly a suboptimal action (see §5.1). After eliminating strategies that prescribe reneging, it is best for an arriving customer who observes n_e to balk.

REMARK 2.2 Unless $\frac{R\mu}{C}$ is an integer, the unique threshold equilibrium strategy is n_e. However, there are other non-threshold equilibria. For example, joining whenever the queue size is *not* n_e, is an equilibrium. This may seem puzzling at first, but if all follow this strategy, then the states corresponding to $n_e + 1$ customers or more are transient. Hence, whatever is prescribed for these states is irrelevant to establishing (or refuting) that a strategy defines an equilibrium. It is true, though, that the threshold strategy n_e is the unique SPE.

REMARK 2.3 If $\frac{R\mu}{C}$ is an integer, then mixing with any probability between joining and balking when observing $n_e - 1$ customers (and otherwise doing as before) is also an equilibrium strategy.

2. The LCFS-PR model

Hassin [66] suggested a way to achieve social optimality without imposing admission fees. This section is devoted to describing this approach and its implications. We adopt Naor's model with two changes:

Assumption 8 changes as follows:

The service discipline is LCFS-PR, that is, a newly arrived customer joins the system and is immediately served, possibly preempting the service of another customer. Preempted customers join a queue where later arrivals get priority over earlier arrivals. When a preempted customer's turn to re-enter service comes, his service is resumed from the point of interruption.

Assumption 9 changes as follows:

At any instant, each customer is allowed to renege at no additional cost and never return. The queue is fully observable at any instant, so that a customer can base his decision on the queue length and on his position in it.

In a FCFS queue, a new customer is placed at the end of the queue, and therefore imposes no negative externalities on customers already in the system. However, this customer may impose negative externalities on *future arrivals*. The essence of the discrepancy between individual and social optimization in Naor's FCFS model lies in the fact that the customer ignores these externalities. Therefore, the individual may join a queue even when his own expected welfare is smaller than the expected reduction in welfare to future customers.

The externalities imposed by a newly arrived customer on those who are presently in the system, are highest if he is assigned to the head of the queue. Under LCFS-PR, every arriving customer is placed at the head of the queue, pushing back those customers who arrived earlier. However, all future arrivals will be placed in front of him and therefore he does not impose any external effects on them.

Hassin observed that LCFS-PR leads to a socially optimal behavior by the customers. The relevant decision that an individual faces is *when to leave* the queue rather than *whether to join* it. By the memoryless property of the exponential distribution it follows that the distributions of the customers' residual service are independent of the queue length and of the amount of service each of them has already received. Since the model assumes homogeneous customers, the waiting customers have identical time and service values as well as identical distributions of residual service time. Therefore, when a customer decides to renege there is no other customer behind him. Since everybody present is served prior to the person at the end of the queue, he imposes no externalities, regardless of his action. In other words, his considerations coincide with those of the society, and hence he will reach the same conclusion of whether or not to renege. In particular, his threshold is n^*. (Note that from the social point of view the order of service is irrelevant, so that the socially optimal threshold is the same under the FCFS and LCFS-PR regimes.) In the next section we use this observation in order to determine n^*.[7]

We now discuss the LCFS-PR model and its implications.

- There is a strategic difficulty associated with the LCFS-PR model. A customer whose service has been preempted is motivated to renege and re-enter the system, pretending to be a new arrival. Such behav-

[7]Remarks 2.1 and 2.2 also apply to the FCFS-PR model. A customer in position n^* knows that the same reasoning that led him not to renege earlier leads everyone in front of him not to renege while in positions 1 to $n^* - 1$. Similarly, the equilibrium is not unique. For example, reneging if and only if there are exactly n^* customers ahead in the queue is also an equilibrium strategy. Yet, the threshold strategy of n^* is the unique SPE.

ior contradicts Assumption 10 in Naor's model and therefore must be prevented administratively.[8]

- The important property of the LCFS-PR model that leads to optimal individual behavior is that the last customer in the queue remains at the end of the queue as long as he is in the system and therefore he imposes no externalities. This property is preserved by any queue discipline with the property that the newly arrived customer is placed anywhere except for at the end of the queue. A particularly appealing policy is to assign a newly arrived customer, whenever the server is busy, to the position before the last. This policy reduces the customer's incentive to renege and re-enter as a new arrival. There is, however, another difficulty associated with this solution. Suppose that customer A is now at the end of the queue while B is just one position ahead of him. If A reneges, then B becomes the last one, and all future arrivals will be positioned ahead of him. Thus B may find it beneficial to offer A a payment so that A doesn't renege. Such side payments must be prevented to preserve optimal behavior. This can be done by concealing the identities of the customers in the queue.

- The solution just proposed has other advantages over LCFS-PR: (i) Preemption may incur some loss of service and this solution is associated with fewer preemptions. (ii) Risk averse customers are worse off under the LCFS-PR discipline than under other queue disciplines like FCFS, since LCFS-PR is associated with a larger waiting time variance.[9] Under the LCFS-PR rule some customers are continuously served without waiting while others wait for long periods of time and finally renege without being served. In particular, in a FCFS queue no customer incurs negative utility (assuming that the utility associated with immediate balking is 0), while this is not the case with LCFS-PR. Assigning new arrivals to the position before last reduces all these drawbacks while maintaining a socially optimal behavior.

- The model is of course a simplified one. However, the qualitative implications are quite general. It is well known that if customers differ by their characteristics (waiting cost, service distribution, service value, etc.) social welfare can be increased by proper assignment

[8]If waiting "at home" is less costly than waiting in the queue it may be socially desirable that a customer returns to the system after balking or reneging. Models with retrials are discussed in §5.

[9]Kingman [87] showed that FCFS (respectively, LCFS-PR) minimizes (respectively, maximizes) the variance of the waiting time.

of priorities (this is the subject of §4). A consequence of the above discussion is that:

– Assigning priorities may be beneficial even when the customers are identical!

Suppose that priorities are assigned randomly or according to some irrelevant basis. The customer at the end of the queue will usually have low priority, and may expect most future arrivals to be placed in front of him. This decreases the externalities he imposes and makes his decision of whether to renege closer to the socially optimal one.

- Olson [136] showed that a LCFS-PR regime can be induced through an appropriate price menu, so that customers receive priority levels based on the amount they paid rather than administratively. Such a pricing system will also achieve social optimality. See also §4.1.3.
- An LCFS-PR discipline induces optimal customers' behavior also in more general observable models. For example, consider an $M/M/s$ system, where the servers may have different service rates. An arriving customer starts service at the *fastest* server. This action may preempt the service of an earlier customer who is then moved to the second fastest server, and so on. A customer at the slowest server may be returned to the queue. It is also possible that the customer at the end of the queue reneges at this stage as a result of the increase in his expected waiting time. As in the single server system (and because of the same reasons), reneging is done in the socially optimal way. Variations of this model with $s = 2$ were analyzed by Xu [177].
- Illustrative descriptions of the LCFS-PR model and its consequences were given by Nalebuff in [132] and Landsburg in [98].

3. Social optimization

In this section we derive the threshold equilibrium strategy under the LCFS-PR regime.[10] As discussed in the previous section, this threshold coincides with n^*, the socially optimal threshold.

Let n be the maximum possible length of the queue, i.e., a customer reneges whenever there are n other customers in front of him. This number includes the one in service. Let f_n be the expected benefit for a customer in position n in the LCFS-PR queue when all (including the customer under consideration) renege from position $n + 1$. Of course,

[10] See Olson [136] for an alternative proof.

f_n is monotone increasing in n and n^* is the largest n such that $f_n \geq 0$. Next we determine the value of f_n in terms of the model's parameters.

LEMMA 2.4

$$f_n = R\frac{1-\rho}{1-\rho^{n+1}} - \frac{C}{\mu(1-\rho)}\left[n-(n+1)\rho\frac{1-\rho^n}{1-\rho^{n+1}}\right]. \tag{2.5}$$

Proof: Under the stated conditions, the probability that the customer eventually receives the benefit of R is the same as the ruin probability in the gambler's ruin problem[11] when the initial asset is n, the goal is $n+1$ and the winning probability in each round equals $p = \frac{\lambda}{\lambda+\mu} = \frac{\rho}{1+\rho}$. Let $q = 1-p$. The ruin probability is

$$\frac{\left(\frac{q}{p}\right)^{n+1} - \left(\frac{q}{p}\right)^n}{\left(\frac{q}{p}\right)^{n+1} - 1} = \frac{1-\rho}{1-\rho^{n+1}}.$$

This value multiplied by R is the positive part of the utility of an individual who is in position n and plans to renege in case he reaches position $n+1$. For the negative part, note that the expected number of rounds until the game is over, that is, the gambler is either ruined or he reaches his goal, is

$$\frac{n}{q-p} - \frac{n+1}{q-p}\frac{1-\left(\frac{q}{p}\right)^n}{1-\left(\frac{q}{p}\right)^{n+1}} = \frac{1+\rho}{1-\rho}\left[n-(n+1)\rho\frac{1-\rho^n}{1-\rho^{n+1}}\right].$$

Multiply this expression by $(\lambda+\mu)^{-1}$ to get the expected time in the system for such a customer and then by C to get the expected waiting cost. □

Based on (2.5), $f_n \geq 0$ if and only if

$$\frac{R\mu}{C} \geq \frac{n(1-\rho)-\rho(1-\rho^n)}{(1-\rho)^2}. \tag{2.6}$$

Denote the right-hand side of (2.6) by $g(n)$ and observe that

$$g(n) \quad = \quad \frac{n(1-\rho)-\rho(1-\rho^n)}{(1-\rho)^2}$$

[11] See, for example, [53].

$$
\begin{aligned}
&= \frac{n - \sum_{i=1}^{n} \rho^i}{1-\rho} \\
&= \sum_{i=1}^{n} i\rho^{n-i} \\
&= \sum_{i=0}^{n-1} (n-i)\rho^i.
\end{aligned}
$$

The function $g(n)$ is unbounded and strictly increasing in n for any fixed $\rho > 0$ whereas $g(0) = 0$. A unique value of n^* therefore exists such that $g(n^* - 1) \leq \frac{R\mu}{C} < g(n^*)$ and this is the maximum number of customers in the system under a welfare maximizing control of the queue.

In Naor's presentation, let $g(\nu) = \frac{\nu(1-\rho)-\rho(1-\rho^\nu)}{(1-\rho)^2}$ be defined over the real line and let ν^* be the unique solution to $g(\nu) = \frac{R\mu}{C}$, then $n^* = \lfloor \nu^* \rfloor$. Also,

$$
\begin{aligned}
g(\nu) - \nu &= \frac{\rho}{1-\rho^2}\left[\nu(1-\rho) - (1-\rho^\nu)\right] \\
&= \frac{\rho(1-\rho)}{1-\rho^2}\left[\nu - (1 + \rho + \cdots + \rho^{\nu-1})\right] \geq 0.
\end{aligned}
$$

Therefore, recalling (2.1),

$$
n^* = \lfloor \nu^* \rfloor \leq \lfloor g(\nu^*) \rfloor = \left\lfloor \frac{R\mu}{C} \right\rfloor = n_e.
$$

In words:

- Individual optimization leads to longer queues than are socially desired.

This result is robust in the sense that it also holds for more general queueing models, as shown in [84, 90, 109, 157, 163, 178, 179].

4. Profit maximization

Suppose now that the server charges an admission fee p but, as opposed to the social point of view where the funds collected are considered to be transfer payments, now they are the server's profits. The model assumes that the fee p is announced and customers base their decision of whether to join the queue on this fee. Thus, a customer who observes i customers in the system enters only if the value R of service is at least as large as the expected full price $p + C\frac{i+1}{\mu}$. Another way to look at this behavior is that customers evaluate service completion only by $R - p$. Given p, the maximal possible length of the queue is, as in (2.4),

$$n = \left\lfloor \frac{(R-p)\mu}{C} \right\rfloor. \tag{2.7}$$

Recall that q_n denotes the probability of observing n customers in the system given that n is also the threshold they adopt, i.e., n is the largest possible number of customers to be present at the same time. Thus, q_n is the *balking probability*. The effective arrival rate is then $\lambda(1 - q_n)$ and the rate of profits is $\lambda(1 - q_n)p$. The server chooses a desired threshold n and sets the maximum price that conforms with this threshold, that is[12]

$$p = R - \frac{Cn}{\mu}. \tag{2.8}$$

This, coupled with (2.2) implies that for a given threshold n, the server's profit is

$$Z_O(n) = \lambda \frac{1-\rho^n}{1-\rho^{n+1}} \left(R - \frac{Cn}{\mu} \right) = \lambda R \frac{1-\rho^n}{1-\rho^{n+1}} \frac{\nu_e - n}{\nu_e}, \tag{2.9}$$

where

$$\nu_e = \frac{R\mu}{C}.$$

A profit-maximizing threshold satisfies the following two conditions: $Z_O(n) > Z_O(n-1)$ and $Z_O(n) \geq Z_O(n+1)$. Substituting in (2.9), the first condition amounts to

$$\frac{1-\rho^n}{1-\rho^{n+1}}(\nu_e - n) > \frac{1-\rho^{n-1}}{1-\rho^n}(\nu_e - n + 1),$$

or

$$(\nu_e - n)\frac{(1-\rho^n)^2 - (1-\rho^{n-1})(1-\rho^{n+1})}{(1-\rho^{n+1})(1-\rho^n)} > \frac{1-\rho^{n-1}}{1-\rho^n}.$$

Assume that $\rho \neq 1$. Note that the fraction in the left-hand side is positive so that both sides can be divided by it without changing the direction of the inequality. This leads to

$$\nu_e - n > \frac{(1-\rho^{n-1})(1-\rho^{n+1})}{\rho^{n-1}(1-\rho)^2}.$$

---

[12]The price p given in (2.8) is such that an arrival who observes $n-1$ customers in the system, where $n = \frac{(R-p)\mu}{C}$, is indifferent between joining (after paying) or balking. Hence, this price is optimal only under the assumption that customers break ties in favor of joining. Otherwise, it is better to charge a little less than p, and formally no optimal price exists.

Substituting $n+1$ for n and reversing the direction of the inequality, the second condition becomes

$$\nu_e - n - 1 \le \frac{(1-\rho^n)(1-\rho^{n+2})}{\rho^n(1-\rho)^2}.$$

These two conditions can be summarized to

$$n + \frac{(1-\rho^{n-1})(1-\rho^{n+1})}{\rho^{n-1}(1-\rho)^2} \le \nu_e < n + 1 + \frac{(1-\rho^n)(1-\rho^{n+2})}{\rho^n(1-\rho)^2}. \quad (2.10)$$

Define a variable $\nu \ge 1$ which depends on ρ for $\rho \ge 0$ through the following relation:

$$\nu_e = \nu + \frac{(1-\rho^{\nu-1})(1-\rho^{\nu+1})}{\rho^{\nu-1}(1-\rho)^2}. \quad (2.11)$$

For given ν_e and ρ, the right-hand side strictly increases from 0 to ∞ with ν.[13] Therefore, there exists a unique solution ν_m to (2.11). Let $n_m = \lfloor \nu_m \rfloor$, then n_m uniquely satisfies the optimality conditions (2.10). The profit-maximizing fee is then $R - n_m \frac{C}{\mu}$, and the corresponding rate of profits is

$$Z_O = \lambda R \frac{1-\rho^{n_m}}{1-\rho^{n_m+1}} \left(1 - \frac{n_m C}{\mu}\right). \quad (2.12)$$

It is immediate that $\nu_e \ge \nu_m$ so that $n_e \ge n_m$. Naor also showed that $n_e \ge n^* \ge n_m$.

We now describe some additional results.

- Naor showed that the profit-maximizing fee is greater than the welfare maximizing fee. Knudsen [90] generalized this result to multi-server queues with nonlinear waiting cost functions. In particular, Naor's result still holds when the benefit from service is decreasing and concave in the time of stay in the system. Simonovits [157] proved a similar result for $GI/M/s$ queues.

- Yechiali [178] showed how to compute the profit-maximizing fee in a $GI/M/1$ queue under two models. In the first, customers react to the fee independently, as in (2.7). In the other, customers are organized and collectively choose a threshold that maximizes their

[13] Naor observed that if $\rho = 1$ no true singularity exists for this function. The function is well behaved there and its value equals $\nu^2 - 1$.

total welfare given the fee, which is considered from their point of view as a real cost (and not a transfer payment, as assumed in Section 3). The server knows the way customers react to the fee and selects the fee in a way that maximizes profits. Let n_m and $\bar{n}_m$ be the thresholds which result under profit maximization when customers react in an individual and in a collective way, respectively. Simonovits [157] proved that in the $M/M/1$ queue $\bar{n}_m \leq n_m$, and conjectured that the same relation holds for a general service distribution.

- Rue and Rosenshine [149] investigated the sensitivity of the thresholds and gains in Naor's model to changes in the arrival rate. The qualitative results are:
  - Under individual optimization, the social welfare is a unimodal function. It increases with λ for small values of λ due to the increased usage of the server, but decreases for larger values due to the increased expected waiting time.
  - The welfare maximizing threshold is a monotone non-increasing function of the arrival rate.
- Hassin [66] also investigated the effect of changes in the arrival rate on the thresholds. As λ grows from 0 to ∞, ν_e is constant, ν^* is non-increasing, and ν_m increases from 1 reaching its maximum value at $\lambda = \mu$, and then decreases back to 1 as $\lambda \to \infty$. For $\lambda < \mu$, when either n^* or n_m changes, the difference between them decreases. However, for $\lambda > \mu$, this difference decreases when n^* changes and increases when n_m changes, since both are decreasing step-functions and $\nu^* > \nu_m$.

  Denote by $\hat{S}_O$ the social welfare *under a profit-maximizing fee* in the observable model. Substituting n_m in (2.3) one gets that

$$\hat{S}_O = \lambda R \left\{ \frac{1-\rho^{n_m}}{1-\rho^{n_m+1}} - \frac{1}{\nu_e}\left[\frac{1}{1-\rho} - \frac{(n_m+1)\rho^{n_m}}{1-\rho^{n_m+1}}\right]\right\}. \qquad (2.13)$$

The functions Z_O and $\hat{S}_O$ are illustrated in Figure 2.1. $\hat{S}_O$ has discontinuities at the values of λ where n_m changes. The jumps are upwards for $\lambda < \mu$ and downwards when $\lambda > \mu$. The functions coincide whenever $n_m = 1$, since in this case the server's profits coincide with the social welfare (the consumer surplus is 0 in both cases). As λ increases, the functions approach $\mu R - C$ which is the net rate of benefit to the customer who is currently served.

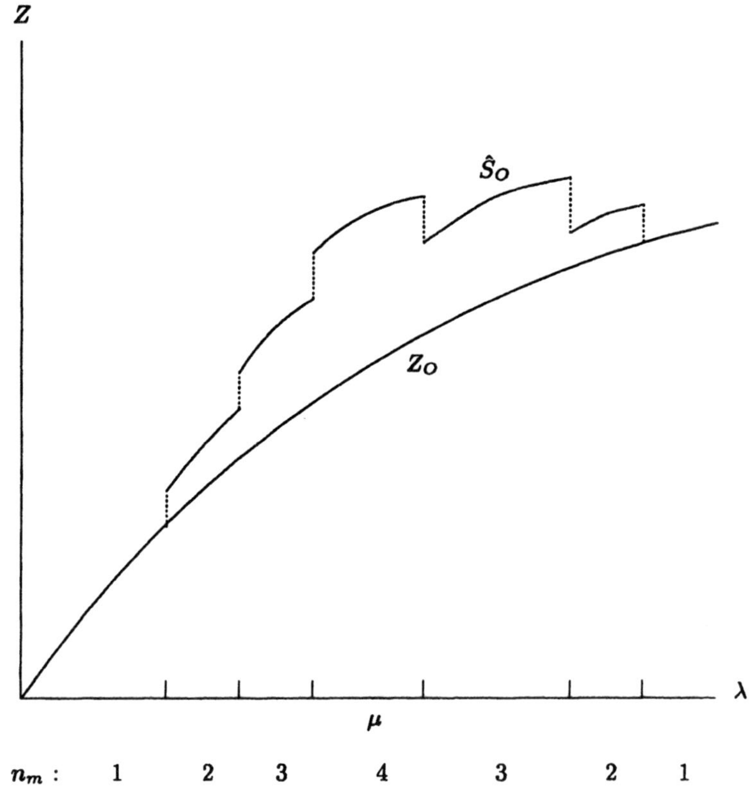

Figure 2.1. Profit and social welfare under a profit-maximizing fee

- In the LCFS-PR queue, the expected welfare of each customer is independent of the queue length at the time of his arrival. A profit-maximizing fee exactly equals the value of expected welfare, and leaves the customers with zero surplus.[14] Under such a fee, every arriving customer joins the system, and his behavior after joining is independent of the fee. In other words, the optimal behavior of the customers is preserved in a LCFS-PR queue even if it is managed by a profit maximizer. This property contrasts the non-optimal behavior induced by a profit maximizer in a FCFS queue.

[14]We assume that indifferent customers join. Otherwise the price should be slightly lower, leaving some positive surplus. See the footnote about (2.8).

5. Heterogeneous customers

Larsen [99] considered a generalization of Naor's model assuming that customers differ by their service values. Suppose that the service value R of an arriving customer is a random variable with a distribution function F. An arriving customer knows his value, but this is his private information and hence it cannot be used by the server to discriminate among customers.

Suppose that an admission fee p is imposed. A customer who observes i customers in the system joins if his service value is at least $p + C\frac{i+1}{\mu}$. Therefore, the joining process when i customers are present, is Poisson with rate

$$\lambda_i = \lambda \bar{F}\left(p + C\frac{i+1}{\mu}\right),$$

where $\bar{F} = 1 - F$. The steady-state probabilities $q_0, q_1, \ldots$ are given by

$$q_i = A_i q_0, \quad i \geq 1$$

and

$$q_0 = \left[\sum_{i=0}^{\infty} A_i\right]^{-1},$$

where for $i \geq 1$

$$A_i = \frac{\lambda}{\mu}\bar{F}\left(p + C\frac{i}{\mu}\right) A_{i-1},$$

and $A_0 = 1$.

The server operates at rate μ as long as the system is not empty. Hence, the average rate of profits to the server is then

$$Z(p) = (1 - q_0)\mu p.$$

The expected social welfare gained from a customer who arrives while the system is in state i is

$$S_i(p) = \bar{F}\left(p + C\frac{i+1}{\mu}\right)\left[E\left(R \,\middle|\, R > p + C\frac{i+1}{\mu}\right) - C\frac{i+1}{\mu}\right].$$

The average welfare contribution per unit of time is

$$S(p) = \lambda \sum_{i=0}^{\infty} q_i S_i(p).$$

This expression may be simplified by substituting

$$\lambda q_i \bar{F}\left(p + C\frac{i+1}{\mu}\right) = \mu q_{i+1},$$

which gives

$$S(p) = \mu \sum_{i=0}^{\infty} q_{i+1} \left[E\left(R \;\middle|\; R > p + C\frac{i+1}{\mu}\right) - C\frac{i+1}{\mu} \right].$$

Larsen conducted a numerical study assuming that R is a continuous uniform random variable, and found that both $Z(p)$ and $S(p)$ are unimodal functions. For the special case where the customers join if and only if the system is empty, Larsen proved that the profit-maximizing fee is greater than or equal to the welfare maximizing fee, as in Naor's model.

In contrast, Edelson and Hildebrand [47] showed that this property does not necessarily hold if customers also differ by their time values. They presented an example with just two types of customers, where various values for the parameters yield a socially optimal admission fee that is smaller, equal, or greater than the profit maximizing fee. Afèche and Mendelson [3] provided conditions under which each of these cases occurs.

The model of Edelson and Hildebrand [47] and Parra-Frutos and Aranda-Gallego [138] assumes two classes of customers with arrival rates λ_i and time values C_i, $i = 1, 2$. This leads to thresholds n_1 and n_2 as in (2.1) such that an i-customer joins if he observes no more than $n_i - 1$ customers in the system.

Schroeter [154] modified Naor's model by assuming that the time value C is uniformly distributed over an interval $[0, C_{\max}]$. This assumption simplifies the derivation of the profit-maximizing price. An individual who observes k customers in the system joins the queue if his time value C is at most $\frac{(R-p)\mu}{k+1}$. Assuming that C_{max} is large enough so that a customer with this time value does not join even when observing an empty queue, the joining process is Poisson with a state dependent rate $\lambda(p,k) = a\frac{(R-p)\mu}{k+1}$, where $a = \frac{\Lambda}{C_{max}}$ and Λ is the potential rate of demand. An analysis of this *proportional balking model* is presented in Page [139]. In particular, the expected rate of service completions is given by

$$\lambda(p) = \mu(1 - e^{-a(R-p)}).$$

The profit maximizer's problem is to select a price p which maximizes $p\lambda(p)$. This price is obtained from the first-order condition

$$1 + ap = e^{a(R-p)}.$$

De Vany [41] considered an observable queue where the demand for service is a function $\Lambda(p)$ of the admission fee. After observing the queue, some customers balk (if the queue size exceeds a *common* threshold), and the resulting joining rate is $\lambda(p)$. We see a problem in this model. The dependence of the potential demand on the admission fee suggests that customers are heterogeneous and we would expect them to have distinct thresholds. Moreover, if there is no cost associated with observing the queue, then every customer who values the service by at least $p + \frac{C}{\mu}$ should arrive first and then decide whether or not to join according to his individual threshold. On the other hand, there may be a cost associated with arriving and observing the queue. In this case, the arrival rate should stabilize at a level which depends on the expected full price (consisting of the arrival cost, admission fee, and waiting cost) and not only on the admission fee. De Vany's main result is that, as in Naor's model, the fee charged by a profit maximizing server is too high and thus the rate by which customers join is too small relative to the socially optimal solution.[15]

Miller and Buckman [128] considered an $M/M/s/s$ model: the first s stands for the number of servers and the second for the maximum number of customers that can stay in the facility at any given time. Hence, there is no queue and an arriving customer who finds all servers busy must balk. The model assumes heterogeneous service values, so that for any given value of service fee, the demand consists of only those customers whose service value exceeds the fee.[16] Miller and Buckman showed that the optimal fee T^* satisfies

$$T^* = \frac{A_s(T^*) - A_{s-1}(T^*)}{\mu}$$

where $A_i(T)$ is the expected social welfare per unit of time when the number of servers is i and the service fee is T. Thus:

- The social welfare maximizing price per expected unit time of service, $T^*\mu$, equals the incremental productivity of the s-th server under the optimal price T^*.[17]

6. Non-FCFS queues without reneging

When the service discipline in an observable queue is not FCFS, the decision of whether or not to join depends not only on the state of the queue but also on the strategy adopted by future arrivals. In Section 2

[15] See [33] for a criticism of the model of [41].

[16] There is no waiting in queue and therefore the value of service is compared to the price and not to a full price.

[17] Note that $A_{s-1}(T^*)$ is not the maximum rate of social gains in the $s-1$ servers model, since the optimal price then is not T^*.

we described the consequences of a LCFS-PR discipline when reneging is allowed. In this section we describe several models where reneging is forbidden. The equilibrium strategies in such models are different in nature from those obtained when reneging is allowed.

6.1. LCFS

Tilt and Balachandran [168] considered a $GI/M/s$ LCFS model (without preemption) where reneging is forbidden, but balking is allowed. Hassin and Haviv [73] considered the same model with emphasis on subgame perfect solutions. Consider first a $GI/M/s/(N+s)$ LCFS model, that is, where the number of customers in the system is at most $N+s$. A customer who observes N customers in the queue balks. Since there is no preemption in the model, a customer who observes a free server joins if and only if $R \geq \frac{C}{\mu}$.

Tilt and Balachandran allowed heterogeneous service and time values, and admission fees that depend on the customer's type. They showed how to compute an SPE when the queue length is bounded. We will consider below the case of homogeneous customers.

Suppose that a customer arrives when all the servers are busy and there are i customers in the queue. Recall that $i = N$ means that the customer must balk, so assume that $0 \leq i < N$. Note that of all possible values of i in this range, the preferred one for a new customer is $i = N - 1$. The reason is that these $N - 1$ customers do not impose any waiting time on the new customer, since they are guaranteed to stay behind him as long as he is in the queue. Moreover, the customer is not concerned with future arrivals, as they are forced to balk as long as the new customer is still in the queue.

The *forced joining strategy* prescribes joining the system as long as this is possible. Let Q_n be the expected queueing time of a customer who observes n, $n = 1, \ldots, N$, vacant positions upon arrival, when all other customers adopt the forced joining strategy. Tilt and Balachandran showed that for $n = 1, \ldots, N$

$$Q_n = \frac{1}{s\mu} \sum_{j=0}^{n-1} \left(\frac{\lambda}{s\mu} \right)^j .$$

Clearly, Q_n is monotone increasing in n: a higher value of n means that more future arrivals are expected to overtake the new customer. The expected net benefit of such a customer, if he joins, is

$$R_n = R - C \left(\frac{1}{\mu} + Q_n \right)$$

which is monotone decreasing in n.

Suppose the model's parameters are set so that for some k, $1 \leq k \leq N-1$, $R_k > 0$ but $R_{k+1} < 0$. Then, the best response for a customer who observes state n (when all others using the forced joining strategy) is to join when $1 \leq n \leq k$ and balk otherwise. Remove now the forced joining assumption. Still by induction we conclude that under an SPE, customers join when $n \leq k$ and balk when $n = k+1$. The fact that customers do not join when $n = k+1$ makes it worthwhile to join when $n = k+2$. Continuing with this line of reasoning we conclude that under the SPE strategy, customers balk if $n = ik+1$ for some $i \leq 1$ and join otherwise.

There are other pure equilibrium strategies. An example for such a strategy when $N > 2k+1$ is as follows: join if and only if the number of customers in the queue is smaller then j for some $j < k$. The strategy does not prescribe an optimal response when the number of customers observed upon arrival is i, $N-k \leq i \leq N-1$: joining is certainly better. Yet, it is still an equilibrium since these states are transient under this strategy.

Consider now the model with $N = \infty$. Assume again that $R_k > 0$ and $R_{k+1} < 0$. The above-mentioned properties are still valid for an equilibrium. However, in this case there are more pure subgame perfect equilibria. An SPE prescribes for some $l \in \{0, \ldots, k-1\}$ joining in all states except for those whose index is $(k+2)i + l$ for some integer $i > 0$. For example, when λ is sufficiently large so that $A_0 > 0$ but $A_1 < 0$, there are two SPE solutions. One prescribes joining if and only if the queue length has an odd value, the other if and only if it is even.

6.2. EPS and random queues

Altman and Shimkin [10] considered a system of observable egalitarian processor sharing (EPS) where reneging is forbidden. Customers decide whether to join the queue after observing the number of customers already there. As in the LCFS model, customers are affected by the strategies adopted by future customers. This is an ATC situation, and Altman and Shimkin showed how to compute the unique (pure or mixed) threshold equilibrium strategy. This strategy is also the unique SPE.[18]

In a similar model, service is granted in random order. In fact, due to the memoryless service process, the two models coincide with respect to the decision problem posed here if the decision about whose service was completed is done randomly among the customers in the queue

[18] An extension where customers differ in their expected service time is considered in Ben-Shahar, Orda and Shimkin [28].

after service completions. When customers *commence* service in random order, the threshold can be computed in a similar way.

A variation of the model, where reneging is allowed, leads to a different model which is analyzed in detail in §5.1.

7. Discounting

Chen and Frank [34] generalized Naor's model assuming that both the customers and the server maximize their expected discounted utility using a common discount rate. Chen and Frank computed the profit and social welfare maximizing pricing schemes for this generalization of Naor's model.[19] We illustrate this approach by describing the customers response to a given admission fee of size p.

Let γ be the discount rate. Consider a customer who joins at time 0 and denote by θ his service completion time. The expected benefit for this customer is

$$E\left[e^{-\gamma\theta}R - p - C\int_0^\theta e^{-\gamma t}dt\right], \tag{2.14}$$

where the expectation E is taken with respect to θ.

Assume that the service times are exponentially distributed with a rate μ. If the arriving customer observes n customers already in the system, then θ follows an Erlang distribution with parameters $n+1$ and μ. In this case, $E[e^{-\gamma\theta}] = \phi^{n+1}$ where

$$\phi = \frac{\mu}{\mu+\gamma}.$$

Therefore, the expression in (2.14) equals

$$\phi^{n+1}\left(R+\frac{C}{\gamma}\right) - \left(p+\frac{C}{\gamma}\right).$$

The customer prefers joining to balking if

$$\phi^{n+1}\left(R+\frac{C}{\gamma}\right) \geq \left(p+\frac{C}{\gamma}\right),$$

or

$$n \leq \log_\phi\left(\frac{p+\frac{C}{\gamma}}{R+\frac{C}{\gamma}}\right) - 1.$$

[19]Naor's model is obtained if the discount rate is 0.

8. State dependent pricing

The discrepancy between the profit-maximizing price and the welfare-maximizing price in Naor's model follows from the monopoly's inability to extract all of the consumer surplus. Therefore, the monopoly's objective differs from the social one. In §3 we will see that this discrepancy doesn't exist in the unobservable version of this model. Chen and Frank [34] observed another case where the profit maximizer's objective coincides with the social one, namely, when the server is able to adjust the price to the state of the system (and the population of customers is homogeneous). They showed that the profit-maximizing pricing scheme is to charge the maximum possible fee that does not deter customers from joining, as long as the queue length is less than a threshold, and to charge a high fee otherwise. All of the consumer surplus then goes to the server whose strategy (of whether to accept or reject a customer) is therefore socially optimal.

Chen and Frank also considered the case where customers have heterogeneous service values. It is assumed that these values are not known to the server and hence they cannot be used to discriminate among customers. In this case, Chen and Frank found that the socially optimal strategy is as in the homogeneous case where the expected service value is used as a common value. Thus, the socially optimal behavior depends on the service distribution only trough its mean value. In particular, there is no loss of generality in assuming an exponential distribution. Moreover, as in the case where the server cannot adjust the fee to the state of the system (Section 5), the profit-maximizing fees tend to be higher than the socially optimal fees.[20]

The profit-maximizing strategy is socially optimal also when customers differ in their attributes, as long as the relevant information is available to the server and can be used to determine the admission fee.

Motivated by applications in the mortgage market, Levy, A and Levy, H [106] considered an $M/M/1$ queue where the server advertises a price p_i (from a given set) whenever there are i customers in the system. It is assumed that there is a demand function D so that the joining rate associated with p_i is $\lambda_i = D(p_i)$. The novel assumption of the model is that a customer who *leaves* at time t pays the price advertised just before t. Thus, upon joining, the customer doesn't know how much he will have to pay for the service. Levy and Levy proved that under

[20]Socially, the exact fees are unimportant, as long as they induce joining and balking in the socially desired states. Hence, by saying that the server charges fees that are too high we mean that the threshold induced by these fees is too low.

the profit-maximizing pricing scheme, $p_{i+1} \geq p_i$ $i = 0, 1, \ldots$, and that the expected profit is higher than in the corresponding system where customers pay the price advertised before their arrival.

We suggest an extension of the model where the joining rates λ_i are determined through an equilibrium mechanism. Suppose that the potential rate of demand is Λ, and that customers differ by their value of service. Consider a known strategy of the server, consisting of the prices $p_0, p_1, \ldots$. The joining rates $\lambda_0, \lambda_1, \ldots$ define an equilibrium if $\frac{\lambda_i}{\Lambda}$ equals the probability that the service value of a random customer is at least the expected full price P_i associated with state i. For given joining rates $\lambda_0, \lambda_1, \ldots$, the expected full prices P_i are computed as follows: let $q_{i,j}$ be the probability that a customer who joins while the system is in state i leaves the system in state j (the state just before he leaves is $j + 1$), then

$$P_i = C\frac{i+1}{\mu} + \sum_{j=0}^{\infty} q_{i,j} p_{j+1}.$$

9. Waiting for the right server

Kumar and Walrand [97] considered a general model of optimal routing which also applies to queues. Consider a $GI/G/s$ system with service rates $\mu_1 \geq \cdots \geq \mu_s$. Whenever a server becomes free it is offered to the waiting customers according to their order in the queue. A customer may either accept the offer or reject it and stay in the queue (in the same position). In the latter case, the service from this server is offered to the next in line. Based on a similar result by Agrawala, Coffman, Garey and Tripathi [5] for a static model with no arrivals, Kumar and Walrand proved that the equilibrium strategy is such that a customer in position i accepts the offer to commence service with server j, if and only if

$$i > \frac{\sum_{k=1}^{j-1} \mu_k}{\mu_j} - (j - 1).$$

Kumar and Walrand proved that under certain assumptions, the equilibrium and socially optimal policies coincide. However, they didn't provide specific examples when the assumptions are satisfied, leaving this issue open.

Another model where customers can delay their decision was considered by Mandelbaum and Yechiali [118]. In their model, all customers join the system unconditionally except for one "smart" customer who is allowed to choose among joining, balking, or waiting outside the system in order to make a decision at the next service completion (choosing again from the three options). Mandelbaum and Yechiali solved the smart customer's problem, however, they did not extend the model for the case where all customers are smart. It is an interesting open question to define such a model and characterize its equilibria.

In another problem of this type, Hlynka, Stanford, Poon, and Wang [80] (see [161] for a generalization) considered a $GI/M/2$ system. Customers, with the possible exception of a single smart customer, follow a strategy of joining the shortest queue. The smart customer has the option of waiting as long as he wishes before deciding which queue to join. During his observation time, new customers may join the system and take their place in front of the smart customer. Again, defining and characterizing equilibrium when all customers are smart is still an open problem for future research.

10. Non-exponential service requirements

Altman and Hassin [9] presented an example for an $M/G/1$ queue without an equilibrium threshold joining strategy. In this example, the service time is 0 with an extremely high probability and 1 with the complement extremely low probability. The information available to an arriving customer is the number of customers in the system. No equilibrium threshold joining strategy exists, since the queue length provides a signal to the arriving customer on the residual service time of the customer currently in service. Altman and Hassin showed that the equilibrium strategy prescribes randomization when there is a customer in service and no queue, joining if the queue size is greater than one but less than some threshold, and balking otherwise.

A similar complication arises when analyzing customers' equilibrium behavior in an $M/G/s$ system, for $s > 1$, where the number of customers in the system is observable but not the time already spent in service by the current customer. In particular, it is not necessarily true that customers select the shortest queue (see, Whitt [175] for the analysis of a similar model).

The socially optimal strategy in a $GI/M/1$ system may include reneging. This may happen when a new arrival is expected soon and the length of the queue is already at its maximum under the optimal strategy. When reneging is not allowed, the optimal joining strategy is of the threshold type. This extension of Naor's model was analyzed by Yechiali [178]. Mendelson and Yechiali [125] analyzed a further refinement of this model allowing "conditional acceptance" of a customer. Specifically, an (n, t)-strategy prescribes joining if the number of customers observed is less than n, and balking otherwise. It also prescribes reneging for the last customer in the queue if there are n customers in the system and t units of time elapsed since the last arrival. The rationale behind this strategy is that for specific inter-arrival distributions, the information that no arrival occurred during t time units indicates that a new arrival is expected soon.[21] Therefore, with high probability, the customer who reneges will soon be replaced, and meanwhile the overall waiting costs are reduced. Mendelson and Yechiali investigated conditions under which the optimal (n, t)-strategy is with $t < \infty$.

11. Related literature

- Naor's model and results have been extended in several papers. Surveys were given by Johansen and Stidham [84], Stidham [164], and Mendelson and Whang [124]. General conditions for optimality of threshold type policies for controlling Poisson input-output systems were given by Hassin and Henig [74].

- Rosenblum [147] considered a model where customers differ by their waiting costs (time value) and have the option of trading positions. The value of service is identical for all customers and each one of them knows the time value of the other customers in the queue. A customer may renege at any time. The resulting queue turns out to be ordered in decreasing value of time. A drawback in this model is (Assumption 9 in [147]) that it assumes that customers ignore possible future profits that may be obtained by trading positions. A model which takes future transactions into account will be more complete.[22].

- Van Ackere and Ninios [1] applied simulation to solve a model where the server can affect the arrival rate by *advertising* the facility. Atkinson refined their results. Let A, n, and q_n be the amount invested in advertising the facility, the equilibrium threshold, and the probability that the length of the queue is n (given that n is the threshold), respectively. Van Ackere and Ninios considered two variations. In one, the potential arrival rate linearly depends on A. In the second, customers also take into consideration the probability that their visit will be unsuccessful, and the potential arrival rate is assumed to be $A(1 - q_n)$.

---

[21] See §5.1 for the concept of increasing hazard rate (IHR).

[22] Holt and Sherman [81] also considered the possibility of resale in their "waiting-line auction" model. See §6.1

Chapter 3

UNOBSERVABLE QUEUES

In the previous chapter we assumed that actions are selected after observing the queue. In this chapter, we present models where the customers do not observe the queue prior to their actions.

1. Identical customers

The properties of the basic unobservable single server queue were discovered by Edelson and Hildebrand [47]. They adopted the first eight assumptions of Naor's observable model as listed in §2.1, and added the following modifications.

Assumption 10 changes as follows:

At the time a customer's need for service arises, he irrevocably either joins the queue or balks. It is not possible for him to observe the queue length before making this decision.

Assumption 8 is relaxed as follows:

The service discipline is strong and work-conserving.[1]

As in the observable model, a customer who joins the queue imposes negative externalities on others and therefore individual optimization leads to excessive congestion unless the queue is regulated.[2] This property will be formally proved below.

[1] For example, the service discipline may be FCFS, LCFS (with or without preemption), EPS, or random order. Under all these disciplines, the expected waiting time is given in (1.4).

[2] In a more general context, Hardin [64] showed that if each individual uses a common resource to maximize his own utility, then the equilibrium has excessive use of the resource.

1.1. Equilibrium

We start by evaluating the customers' behavior in equilibrium when an admission fee of size p is imposed and the potential arrival rate is Λ. There are two pure strategies available for a customer: to join the queue or not to join. A pure or mixed strategy can be described by a fraction q, $0 \leq q \leq 1$, which is the probability of joining. Given that the admission fee is p, we denote the equilibrium probability of joining by $q_e(p)$, and the corresponding equilibrium (effective) arrival rate by $\lambda_e(p)$. Of course, $\lambda_e(p) = q_e(p)\Lambda < \mu$. We denote the expected waiting time when the (effective) arrival rate is $\lambda < \mu$ by $W(\lambda) = \frac{1}{\mu - \lambda}$. (For $\lambda \geq \mu$, define $W(\lambda) = \infty$.) This function is continuous and monotone increasing. The net benefit for a customer who joins the queue is the value of service, R, minus the full price, $p + CW(\lambda)$. We distinguish three cases:

- $p + CW(0) \geq R$. In this case, even if no other customer joins, the net benefit of a customer who joins is non-positive. Therefore, the strategy of joining with probability $q_e(p) = 0$ is an equilibrium strategy and no other equilibrium is possible. Moreover, in this case, not joining is a dominant strategy.

- $p+CW(\Lambda) \leq R$. In this case, even if all potential customers join, they all enjoy a non-negative benefit. Therefore, the strategy of joining with probability $q_e(p) = 1$ is an equilibrium strategy and no other equilibrium is possible. Moreover, in this case, joining is a dominant strategy.

- $p + CW(0) < R < p + CW(\Lambda)$. In this case, if $q_e(p) = 1$ then a customer who joins suffers a negative benefit. Hence, this cannot be an equilibrium strategy. Likewise, if $q_e(p) = 0$, a customer who joins gets a positive benefit, more than by balking. Hence, this too cannot be an equilibrium. Therefore, there exists a unique equilibrium strategy where $q_e = \frac{\lambda_e(p)}{\Lambda}$ and where $\lambda_e(p)$ solves $CW(\lambda_e(p)) = R - p$.

Substituting $W(\lambda) = \frac{1}{\mu - \lambda}$, we obtain the expressions given in Table 3.1.

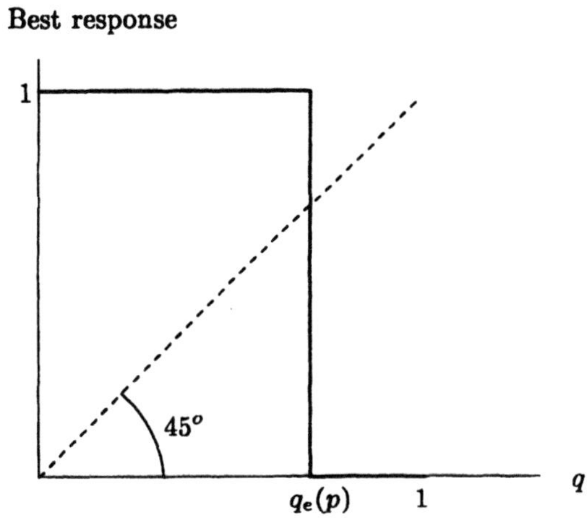

Figure 3.1. The best response vs. the joining probability

Case	$\lambda_e(p)$	$q_e(p)$	$W(\lambda_e(p))$
$\Lambda \leq \mu - \frac{C}{R-p}$	Λ	1	$\frac{1}{\mu-\Lambda}$
$0 \leq \mu - \frac{C}{R-p} \leq \Lambda$	$\mu - \frac{C}{R-p}$	$\frac{\mu-\frac{C}{R-p}}{\Lambda}$	$\frac{R-p}{C}$
$\mu - \frac{C}{R-p} < 0$	0	0	$\frac{1}{\mu}$

Table 3.1. The equilibrium strategy

REMARK 3.1 Suppose that the probability of joining is q: if $q < q_e(p)$ then the unique best response is 1, if $q = q_e(p)$ then any strategy between 0 and 1 is a best response, and if $q > q_e(p)$ then the unique best response is 0. Since the best response is a monotone non-increasing function of the strategy, the model is of the ATC type. Figure 3.1 depicts the best response function and the equilibrium point.

1.2. Social optimization

We now turn our attention to social optimization. Let the socially optimal joining probability be q^*, and let the socially optimal joining rate be λ^* where $\lambda^* = q^*\Lambda$. Then,

$$\lambda^* = \arg \max_{0\le\lambda\le\Lambda} \{\lambda[R - CW(\lambda)]\}.$$

Since $W(\lambda)$ is strictly convex, the function to be maximized is strictly concave and has a unique maximum. Substituting $W(\lambda) = \frac{1}{\mu-\lambda}$ we get that the solution

$$\mu - \sqrt{\frac{C\mu}{R}} = \arg \max_{0\le\lambda<\mu} \left\{\lambda R - C\lambda\frac{1}{\mu - \lambda}\right\}$$

is optimal as long as it is in $[0, \Lambda]$. The fact that the solution is nonnegative follows from the assumption that $R\mu \ge C$. Thus, if $\Lambda \ge \mu - \sqrt{\frac{C\mu}{R}}$ then $\lambda^* = \mu - \sqrt{\frac{C\mu}{R}}$. Otherwise, $\lambda^* = \Lambda$. Let S_U [3] be the social welfare under the optimal arrival rate λ^*. Table 3.2 summarizes the optimal joining strategy.

Case	λ^*	q^*	$W(\lambda^*)$	S_U
$\Lambda \ge \mu - \sqrt{\frac{C\mu}{R}}$	$\mu - \sqrt{\frac{C\mu}{R}}$	$\frac{\mu-\sqrt{\frac{C\mu}{R}}}{\Lambda}$	$\sqrt{\frac{R}{C\mu}}$	$\left(\sqrt{R\mu} - \sqrt{C}\right)^2$
$\Lambda \le \mu - \sqrt{\frac{C\mu}{R}}$	Λ	1	$\frac{1}{\mu-\Lambda}$	$\Lambda\left(R - \frac{C}{\mu-\Lambda}\right)$

Table 3.2. The socially optimal strategy

It follows from the assumption $R\mu \ge C$ that $\lambda_e(0) \le \lambda^*$. Thus, as in the case of observable queues, individual optimization leads to queues that are longer than are socially desired. This gap can be corrected by imposing an appropriate admission fee, as discussed in the next section.

Balachandran and Srinidhi [23] observed that if $\lambda_e(0) < \Lambda$ then

$$\left(1 - \frac{\lambda^*}{\mu}\right)^2 = 1 - \frac{\lambda_e(0)}{\mu}.$$

[3]The subscript U stands for *unobservable.*

REMARK 3.2 Assume that $\lambda^* < \Lambda$. Consider a tagged customer who is given the lowest possible priority so that he is served only when there are no other customers in the system. In particular, his service may be preempted (and resumed later from the point of interruption). The change in the tagged customer's priority has no effect on social welfare. However, with this change, the tagged customer imposes no externalities. By (1.6) his expected waiting cost when the arrival rate equals λ^* is $\frac{C}{\mu}\left(1-\frac{\lambda^*}{\mu}\right)^{-2}$, which by Table 3.2 equals R. Hence, if $\lambda = \lambda^*$, the tagged customer is indifferent between joining and not; if $\lambda < \lambda^*$, he prefers joining; and if $\lambda > \lambda^*$, his choice is not to join. Thus, as expected, when $\lambda = \lambda^*$ the tagged customer who imposes no externalities behaves in the socially desired way. This principle will be used in §4.5 to present a decentralized way for optimally controlling an unobservable queue.

REMARK 3.3 We assumed that the service duration follows an exponential distribution. However, unlike in the observable case, the same qualitative results are obtained for any service distribution. Balachandran and Srinidhi [23] examined the sensitivity of the solution to uncertainty, as reflected by the second moment $\overline{x^2}$ of the service time distribution. They concluded that both λ^* and $\lambda_e(0)$, as well as the ratio $\frac{\lambda^*}{\lambda_e(0)}$, are monotone decreasing in $\overline{x^2}$. The latter property means that:

- The need to control the queue becomes more critical as the uncertainty measured by the variance of the service requirement, increases.

1.3. Profit maximization

We consider now a monopolistic server that sets a profit-maximizing admission fee p_m. A monopoly does not leave a positive customer surplus, since in such a case the admission fee can be increased without reducing the arrival rate. Therefore, $p_m + CW(\lambda) = R$. The monopoly's problem is to maximize λp_m subject to $0 \le \lambda \le \Lambda$ and

$$p_m = R - CW(\lambda). \tag{3.1}$$

Recall that the social objective is to maximize the total welfare of the server and the customers, which is $\lambda p + \lambda[R - CW(\lambda) - p]$. The term λp cancels, reflecting the assumption that social utility is additive so that from social point of view the admission fees are merely transfer payments that have no effect on social welfare. Hence, the social problem is to maximize $\lambda[R - CW(\lambda)]$ subject to $0 \le \lambda \le \Lambda$. By (3.1):

- The objectives of a profit maximizer and the society coincide.

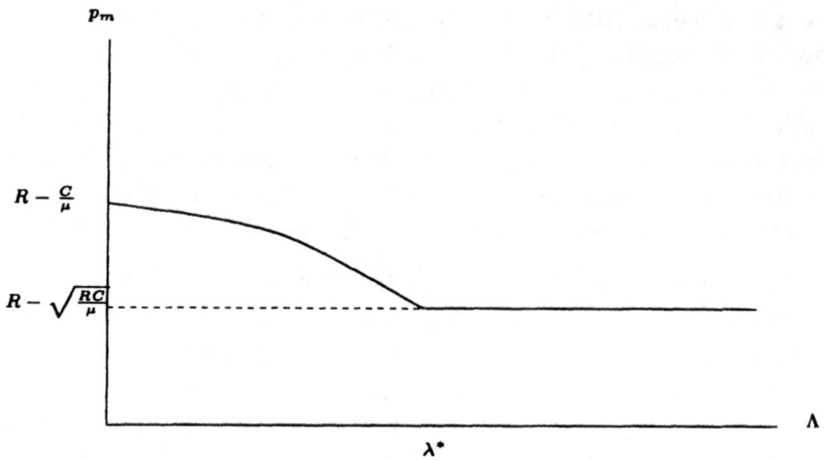

Figure 3.2. Monopoly prices vs. rate of arrival

The socially optimal arrival rate λ^* can be induced by an appropriate admission fee, which also maximizes total profit. When $\lambda^* < \Lambda$ this fee equals

$$p_m = p^* = R - CW(\lambda^*) = R - \sqrt{\frac{CR}{\mu}}.$$

When $\lambda^* = \Lambda$, the profit maximizer chooses the maximum fee which induces this rate, that is, $p_m = R - \frac{C}{\mu - \Lambda}$. A social planner would choose this fee, or any smaller fee, since any such choice induces the same optimal arrival rate, Λ.

Chen and Frank [33] observed that:

- p_m is *monotone non-increasing* in Λ (see Figure 3.2).

Thus, an increase in demand may lead to a reduction in price! This may seem counterintuitive, but can be explained as follows: when the arrival rate increases the expected waiting time increases, and customers are inclined to pay less for service. In other words:

- Since the quality of the goods depends inversely on the demand, the price needs to be reduced when demand increases.

Edelson and Hildebrand also discussed an extension of their model in which the server imposes a *two-part tariff*. For a given *inspection fee* a customer may choose to inspect the queue length and then choose between balking and paying an *admission fee* to join the queue. For any given admission fee, the profit-maximizing inspection fee coincides with the customer's expected gain from inspecting the queue. Thus again, all of the customer's surplus goes to the server and the server chooses socially optimal fees.

REMARK 3.4 The model does not allow reneging. However, this assumption is redundant if we assume that the queue remains unobservable also after joining. This is because the expected residual waiting time is non-increasing with the time already spent in the queue (see §5.2). In an $M/M/1$ queue, due to the memoryless property of the waiting time, this value is constant, whereas when reneging is exercised by a positive fraction of the customers, it is strictly decreasing with the time in the queue.

REMARK 3.5 Joining with probability $q_e(p)$ is an equilibrium strategy also in a LCFS-PR *observable* queue without reneging. The reason is that the length of the queue is unimportant to the new arrival. However, as shown by Tilt and Balachandran [168] and Hassin and Haviv [73], this is not the unique SPE.

2. Observable vs. unobservable queues

Hassin [66] compared social welfare and profit maximization in the observable and unobservable models. Let S_U and S_O denote the social welfare under a social welfare-maximizing policy in the unobservable and observable models, respectively. Similarly, Z_U and Z_O denote the profit under a profit-maximizing admission fee. Recall from Section 1.3 that

$$Z_U = S_U.$$

Let $\hat{S}_O$ denote the social welfare in the observable model *when a profit-maximizing admission fee is charged.* Note that there is no need to define the respective variable $\hat{S}_U$ since the profit-maximizing fee in the unobservable model is the same as the welfare-maximizing fee.

If the queue is controlled by a social welfare maximizing fee, an observable queue would have a higher welfare than the corresponding unobservable queue. In the observable case, a customer would enter only when it is socially desirable to do so, whereas in the unobservable case only the probability that a customer joins is controlled, and it is still possible for customers to enter when the queue is too long or to balk when it is too short. Hence

$$S_U \leq S_O.$$

Hassin reached the following conclusions:

- If $R\mu \leq 2C$ then $Z_U < Z_O$ for all $\Lambda > 0$. This follows by comparing $Z_U = S_U$ from Table 3.2 with Z_O in (2.12). Hence the profit maximizer prefers to reveal the queue length to the customers if this can be done without cost.

- If $R\mu > 2C$ then a unique potential arrival rate Λ^Z exists such that $Z_U > Z_O$ for $\Lambda < \Lambda^Z$, and $Z_U < Z_O$ for $\Lambda > \Lambda^Z$. Thus, when $\Lambda < \Lambda^Z$ the profit maximizer prefers to conceal the queue length from the customers whereas when $\Lambda > \Lambda^Z$ he prefers to disclose this information.

- The same properties hold in respect to $\hat{S}_O$ and S_U for a different threshold value, Λ^S. This follows by comparing S_U from Table 3.2 with $\hat{S}_O$ in (2.13). Thus, when $\Lambda < \Lambda^S$ it is socially preferred that the profit maximizer will be unable to inform the customers on the queue length, whereas when $\Lambda > \Lambda^S$ an observable queue gives (under profit maximization) a higher value of social welfare than the observable queue.

- Figure 3.3 illustrates the case $R\mu > 2C$, and shows that $\Lambda^Z > \Lambda^S$.

- For arrival rates $\Lambda^S < \Lambda < \Lambda^Z$, the profit maximizer prefers to conceal the queue length but social welfare would increase if the server could be induced to disclose this information.

- It is never socially worthwhile to induce the profit maximizer to conceal the queue length (when he does not voluntarily do so).[4]

Chen and Frank [33] compared the effective arrival rates in the observable and unobservable models assuming a *fixed admission fee*. Let λ_O and λ_U be the equilibrium arrival rates in the observable and unobservable models, respectively. Clearly $\lambda_O < \Lambda$, and if Λ is small then $\lambda_U = \Lambda$ and therefore in this case $\lambda_U > \lambda_O$. On the other hand, if Λ is very large then in the observable model the server is almost always active and $\lambda_O \approx \mu$. In the unobservable system, customers still choose a sufficiently small joining rate to avoid high congestion and therefore λ_U is significantly smaller than μ. In fact, as can be seen from Table

[4]Schroeter [154] made a similar comparison in a model where time values are uniformly distributed over an interval $[0, C_{max}]$. He found that for sufficiently large arrival rates, the server would never have an incentive to conceal the queue length, nor would it be in the interest of society to do so. This result conforms with Hassin's conclusions for the case of common time values. Hassin found, however, that for smaller values of λ the outcome is different.

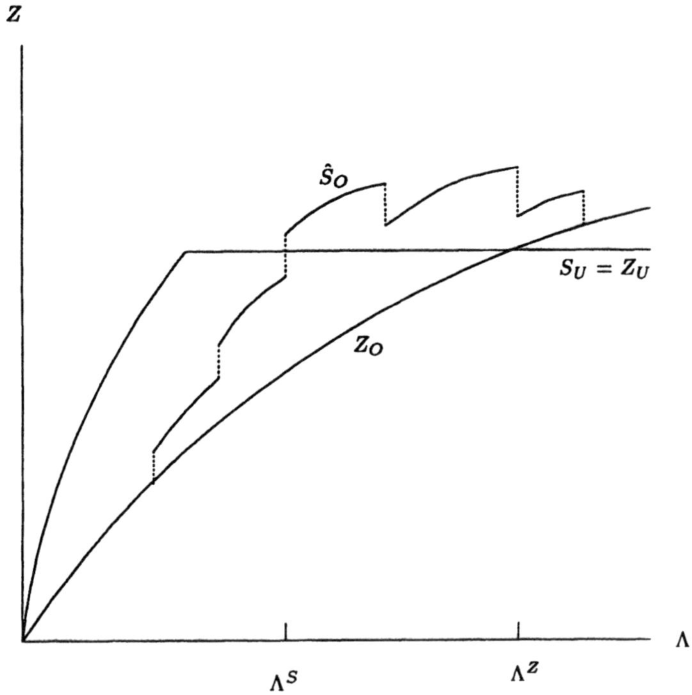

Figure 3.3. Profits and welfare in observable and unobservable queues

3.1, when Λ is large μ and λ_U differ by the constant $\frac{C}{R-p}$. Thus, in this case $\lambda_U < \lambda_O$. Furthermore, Chen and Frank proved that the difference $\lambda_O - \lambda_U$ monotonically increases with Λ, so that there exists a unique critical value Λ^* for which the two rates coincide. In particular, for $\Lambda < \Lambda^*$, $\lambda_U > \lambda_O$, and for $\Lambda > \Lambda^*$, $\lambda_U < \lambda_O$.

Larsen [99] conducted numerical experiments comparing the profits and social welfare in the observable and unobservable models, while assuming that the service values were uniformly distributed in the population. Larsen found that in most cases the values obtained in the observable case were higher. When the opposite result was obtained, the difference was quite small.

3. Heterogeneous service values

Littlechild [110] extended the model of Edelson and Hildebrand assuming that customers have different service values. For $0 \le \lambda < \infty$, let $V(\lambda)$ denote the expected total value of completions of service (per unit of time) corresponding to an (effective) arrival rate λ. The function V

represents a continuous distribution of service values in the population of potential customers. In equilibrium, there is a threshold such that only customers with service values above this threshold arrive to obtain service. This explains why V is assumed to be concave: $V'(\lambda)$ represents the (positive but decreasing) marginal social gain from a customer who joins the system.[5]

The results obtained by Littlechild for the $M/M/1$ queue were generalized to a $GI/G/s$ model and extended further by Mendelson [123], as we now describe.[6]

As before, let λ^* be the socially optimal arrival rate. Then,

$$\lambda^* = \arg\max_{\lambda \geq 0} \{V(\lambda) - \lambda C W(\lambda)\}.$$

An optimal rate λ^* satisfies (and can be determined by) the first-order condition

$$V'(\lambda^*) = CW(\lambda^*) + \lambda^* C W'(\lambda^*). \tag{3.2}$$

The first term in the right-hand side is the customer's cost due to his own waiting time. The second is an *externality cost* he imposes on others, defined as the rate by which waiting costs for the other customers increase when the arrival rate increases by an infinitesimal amount. Thus, from the social point of view, customers should join as long as the marginal increase in social utility due to their service is larger than the marginal increase in costs due to congestion.

Under individual optimization, customers ignore the externalities that they impose and the equilibrium rate λ_e satisfies $V'(\lambda_e) = CW(\lambda_e)$. Comparing this equation with (3.2) we conclude that $\lambda_e > \lambda^*$.

This discrepancy can be corrected by imposing an admission fee p^*, so that λ^* becomes the new equilibrium arrival rate, i.e., λ^* satisfies

$$V'(\lambda^*) = p^* + CW(\lambda^*). \tag{3.3}$$

[5]Another interpretation of the model is as follows. Consider a potential arrival rate of Λ. A customer's value of service is a continuous random variable with density function f, with cumulative distribution function F. Let $\bar{F} = 1 - F$. In equilibrium, the arrival rate is such that the value of service for the marginal customer equals the expected full price associated with joining the queue. Thus, only those who value service by more than a threshold, say a, join. This leads to an arrival rate of $\Lambda\bar{F}(a)$. The social optimal threshold a^* satisfies

$$a^* = \arg\max_{a \geq 0} \left[\Lambda \int_{x=a}^{\infty} x f(x)\,dx - C\Lambda\bar{F}(a) W(\Lambda\bar{F}(a))\right].$$

On the other hand, in equilibrium, customers join the queue if and only if they value service by more than a_e, where a_e uniquely solves $a_e = CW(\Lambda\bar{F}(a_e))$.

[6]De Vany and Saving [42] also considered a similar model.

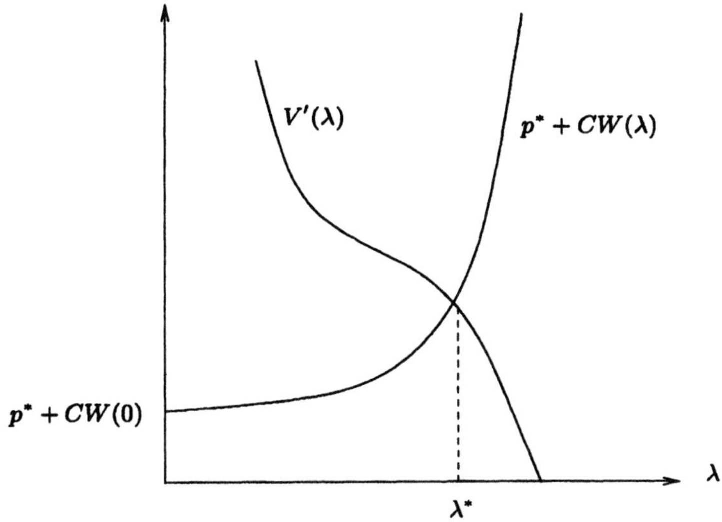

Figure 3.4. Supply and demand curves

The roles of the marginal gain $V'(\lambda)$ and the full price $p + CW(\lambda)$ are similar to those of supply and demand curves and the situation is depicted in Figure 3.4.

Combining (3.2) and (3.3),

$$p^* = \lambda^* CW'(\lambda^*). \tag{3.4}$$

Hence:

- The optimal admission fee p^* coincides with the externalities that an individual who joins imposes on the others, when the arrival rate is λ^*.[7]

Consider now the problem of a profit-maximizing monopoly. The problem is to determine an admission fee p_m such that the resulting equilibrium arrival rate λ_m maximizes $\lambda p = \lambda[V'(\lambda) - CW(\lambda)]$. The first-order conditions are

$$V'(\lambda_m) + \lambda_m V''(\lambda_m) = CW(\lambda_m) + \lambda_m CW'(\lambda_m). \tag{3.5}$$

Observe that (3.2) and (3.5) are identical except for the presence of the negative term $\lambda_m V''(\lambda_m)$ in the left-hand side of (3.5). Therefore,

$$CW(\lambda_m)+\lambda_m CW'(\lambda_m)-V'(\lambda_m) < CW(\lambda^*)+\lambda^* CW'(\lambda^*)-V'(\lambda^*) = 0.$$

[7] This phenomenon exists in other models of congestion. For a review in the context of road pricing, see [130].

By the convexity of W and the concavity of V,

$$\frac{d}{d\lambda}\left[CW(\lambda) + \lambda CW'(\lambda) - V'(\lambda)\right] = 2CW'(\lambda) + \lambda CW''(\lambda) - V''(\lambda) > 0.$$

Therefore:

- $\lambda_m < \lambda^*$.
- $p_m = V'(\lambda_m) - CW(\lambda_m) > V'(\lambda^*) - CW(\lambda^*) = p^*$.

These qualitative results are similar to those obtained by Naor [133] for an observable queue and identical customers. However, in the unobservable case they were obtained under more general conditions.

Larsen [99] analyzed the sensitivity of the maximum profit and welfare to an increased heterogeneity in the value of service. Larsen assumed that the value of service in the population is a continuous random variable with uniform distribution over an interval $[a - \Delta, a + \Delta]$. Larsen showed that an increase in Δ always leads to an increase in the social welfare obtained under a welfare-maximizing admission fee. This result is expected since when Δ increases, those who join value the service more highly whereas those whose value of service is low do not join in any case. On the other hand, Larsen found that an increase in Δ may lead to either an increase or a decrease in the profit obtained under a profit-maximizing fee.

4. Heterogeneous service values and time costs

Suppose that customers belong to one out of m classes where class i is characterized by service and time values R_i and C_i, respectively. For simplicity, assume that the ratios $\frac{R_i}{C_i}$ are distinct and $\frac{R_1}{C_1} > \frac{R_2}{C_2} > \cdots > \frac{R_m}{C_m}$. Let Λ_i be the potential arrival rate of class i, $i = 1, \ldots, m$.

4.1. Equilibrium

Balachandran and Schaefer [21, 22] characterized the equilibrium solution: for some $1 \le r \le m$, classes $1, \ldots, r-1$ arrive at their maximum rates Λ_i, r-customers split between joining and balking, and the other classes fully balk. Splitting of class r is possible if $R_r = C_r W(\lambda)$ where λ is the resulting aggregate arrival rate, so that r-customers are indifferent between joining and balking. To summarize, $\lambda = \sum_{i=1}^{r-1} \Lambda_i + \lambda_r$ where r and λ_r are determined by $R_r = C_r W(\lambda)$. Of course, $R_i > C_i W(\lambda)$ for $i = 1, \ldots, r-1$ and $R_i < C_i W(\lambda)$ for $i = r+1, \ldots, m$.[8]

[8]Section 3.2 of [22] contains further properties of the equilibrium in this case.

Suppose now that the potential arrival rates $\Lambda_1, ..., \Lambda_m$ are infinite. If $R_i - C_iW(\lambda) > 0$ then more members of this class join the queue. Consequently, in equilibrium only 1-customers join the queue, and λ_1 satisfies $R_1 = C_1W(\lambda_1)$. This is known as the *class dominance* phenomenon, which holds also under social optimization (see below) but not necessarily with the domination of the same class.

4.2. Social optimization

Balachandran and Schaefer [21, 22] observed (assuming unbounded potential arrival rates) that social welfare is also maximized when all but a single class fully balk. The reasoning is that for any given aggregate arrival rate λ, the social goal is to maximize $\sum_{i=1}^{m} \lambda_i[R_i - C_iW(\lambda)]$ subject to $\sum_{i=1}^{m} \lambda_i = \lambda$ and $\lambda_i \geq 0$ $i = 1\ldots, m$. This "continuous knapsack problem" is optimized by admitting only one class j which maximizes $R_i - C_iW(\lambda)$ over $1 \leq i \leq m$, and setting $\lambda_j = \lambda$ and $\lambda_i = 0$ for $i \neq j$. Assume now an $M/M/1$ system. Suppose that it is socially optimal that class j dominates the system, so that $\lambda_i = 0$ for every $i \neq j$. Then, λ_j is determined by Table 3.2, and the social welfare is $\sqrt{R_j\mu} - \sqrt{C_j}$. It follows that the social problem is to solve

$$\max_{1 \leq i \leq m} \left(\sqrt{R_i\mu} - \sqrt{C_i}\right)$$

and set for the maximizing class j, $\lambda_j = \mu - \sqrt{\frac{C_j\mu}{R_j}}$ (and $\lambda_i = 0$ for $i \neq j$). This solution may be different from the equilibrium solution, which admits a class with a maximum value of the ratio $\frac{R_i}{C_i}$. Hence, even in the $M/M/1$ case, the class that uses the facility in equilibrium may differ from the socially desired class (and then, given this class, the arrival rate exceeds the socially optimal one).

4.3. Class decision

Balachandran and Schaefer [22] and Gibbens and Kelly [56] considered a class decision model with m classes. In their model, the utility of class j is $\lambda_j[R_j - C_jW(\lambda)]$, where λ_j denotes the arrival rate of j-customers, $j = 1, \ldots, m$, and $\lambda = \sum_{i=1}^{m} \lambda_i$. In equilibrium, $R_j - C_jW(\lambda) - \lambda_jC_jW'(\lambda)$ is 0 if $\lambda_j > 0$, and non-positive if $\lambda_j = 0$. Thus,

$$\lambda_j = \left[\frac{R_j - C_jW(\lambda)}{C_jW'(\lambda)}\right]^+, \quad j = 1, \ldots, m. \tag{3.6}$$

Summation gives

$$\lambda = \sum_{j=1}^{m} \left[\frac{R_j - C_j W(\lambda)}{C_j W'(\lambda)} \right]^+ .$$

The right-hand side is monotone decreasing in λ (recall that W is convex) and hence a unique solution λ exists. Then, using (3.6), λ_j, $1 \leq j \leq m$, can be determined.

5. Customers know their demand

Consider the model of Section 1 but with one change: customers base their decision of whether to join the queue on their service time. We assume that the service time is private information of the customer. We consider equilibrium threshold strategies under two service disciplines: FCFS, and egalitarian processor sharing (EPS). In both cases, the equilibrium behavior is based on a threshold, x_e, such that only customers with service requirement $t \leq x_e$ join. By the same reasoning as in Section 1, the socially optimal threshold, x^* is smaller than x_e.

Assume that the service time is a continuous random variable with distribution and density functions G and g, respectively, and that Λ is the potential arrival rate. Given that only customers with service duration of at most x join, the arrival rate is $\lambda(x) = \Lambda G(x)$.

5.1. FCFS

Suppose that only customers with a service requirement of at most x join. Then, the density function of service requirements among customers who join is $\frac{g(y)}{G(x)}$ for $0 \leq y \leq x$ (and 0 elsewhere), and the expected service requirement among them is

$$m(x) = \frac{1}{G(x)} \int_{y=0}^{x} y g(y)\, dy.$$

Let

$$\rho(x) = \Lambda G(x) m(x) = \Lambda \int_{y=0}^{x} y g(y)\, dy \tag{3.7}$$

be the effective utilization factor of the system. By the Khintchine-Pollaczek formula (1.7), the expected queueing time for those who join is[9]

$$W_q(x) = \frac{\Lambda \int_{y=0}^{x} y^2 g(y)\, dy}{2(1 - \rho(x))}.$$

[9]Assuming an $M/M/1$ model rather than $M/G/1$ does not lead to a simpler analysis: the service distribution of those who join is never exponential.

The equilibrium threshold x_e is defined by

$$C(x_e + W_q(x_e)) = R,$$

whereas the social optimal threshold x^* is

$$x^* = \arg\max_x \{\Lambda G(x)(R - C[m(x) + W_q(x)])\}.$$

5.2. EPS

The EPS model was solved by Haviv [75]. In an $M/G/1$ EPS queue with a utilization factor ρ, the expected time in the system for a customer with service requirement t is (see, for example, [148] p. 174)

$$W_t = \frac{t}{1-\rho}.$$

When the threshold strategy x is applied, the expected time in the system for a customer with service requirement t ($t > x$ is possible here) is

$$W_t(x) = \frac{t}{1-\rho(x)},$$

where $\rho(x)$ is as defined in (3.7). The equilibrium threshold x_e is given by

$$\frac{Cx_e}{1-\rho(x_e)} = R.$$

Using (1.3), the expected number of customers in the system, under the threshold x, is

$$\frac{\rho(x)}{1-\rho(x)}.$$

Hence,

$$x^* = \arg\max_x \left[\Lambda G(x)R - \frac{C\rho(x)}{1-\rho(x)}\right],$$

and by the first-order conditions, x^* satisfies

$$\frac{Cx^*}{[1-\rho(x^*)]^2} = R.$$

It is possible to regulate the system by imposing an admission fee of T such that

$$R - T = \frac{Cx^*}{1-\rho(x^*)},$$

or a fee of p per unit of service time such that $px^* = T$, or a fee t per unit of time in the system such that

$$R = \frac{(C+t)x^*}{1-\rho(x^*)}.$$

5.3. Shortest service first

Suppose now that the service requirement of an arrival is also known to the server, and that the service regime is such that customers with shorter service requirements receive preemptive priority over customers with longer requirements. The same rule applies to the order in which preempted customers return to service. Note that priority levels are based on *original* requirements and not on the *residual* requirements which change while in the system. Service is resumed from the point where it was interrupted.

Customers know their demand and need to decide irrevocably whether or not to join. Consider an equilibrium strategy of the threshold type: for some x_e, all those with demand $x \leq x_e$ join, whereas the others balk. Denote by x^* the socially optimal threshold. We observe that the customer with the longest service requirement who joins imposes no externalities on the others. Therefore, his objective coincides with the social objective, and we conclude that he joins if and only if this is socially desired. Thus, we conclude that $x_e = x^*$.

We next show how to compute this common threshold. Let T_x denote the expected waiting time of a customer whose service time is x, when the threshold strategy x is used by all. Then (see, for example [89] p. 124),

$$T_x = \frac{x(1-\rho(x)) + \Lambda \int_{y=0}^{x} g(y)y\,dy}{[1-\rho(x)]^2}$$

T_{x_e} uniquely satisfies

$$CT_{x_e} = R.$$

6. Finite buffer

Lin and Ross [108] considered a queueing system with a waiting area (or *buffer*) of bounded size. In their model, arrivals are forced to balk when the buffer is full. Since admission to the queue is not guaranteed, we refer to an arrival as a *trial*.[10]

To illustrate such models, Lin and Ross considered an $M/M/1/1$ system, namely a single server system with no waiting room. The service value is R, and without loss of generality assume that the time value is $C = 0$.[11] There is also a cost $T < R$ associated with each trial. This is a real cost, not a transfer payment. The potential arrival rate is denoted by Λ, and let $\rho = \frac{\Lambda}{\mu}$. The customer's problem in this model is: to try or not to try.

---

[10] A model where rejected customers retry later is discussed in §6.4.

[11] Otherwise, just replace R with $R - \frac{C}{\mu}$ and C with 0.

A strategy is characterized by the probability p that a customer tries. The effective trial rate is then $\lambda = p\Lambda$ and the server is idle with probability $\frac{\mu}{\mu+p\Lambda}$. The (expected) payoff for one who tries is therefore

$$\frac{R\mu}{\mu + p\Lambda} - T. \tag{3.8}$$

Clearly, this is an ATC situation and therefore there exists a unique equilibrium. By (3.8), if $\Lambda < \mu\frac{R-T}{T}$, then regardless of what the others do, one's best action is to try (in other words, trying is a dominant strategy). Otherwise, no dominant strategy exists. If $p = \mu\frac{R-T}{T\Lambda}$ then a customer is indifferent between trying or not. Hence, trying with this probability is the equilibrium strategy. Denote the equilibrium trial probability by p_e, then

$$p_e = \begin{cases} 1 & \rho \le \frac{R-T}{T} \\ \frac{R-T}{T\rho} & \rho > \frac{R-T}{T} \end{cases}.$$

The socially optimal trial probability is defined by

$$p^* = \arg\max_p \; p\left(\frac{R\mu}{\mu + p\Lambda} - T\right).$$

Hence,

$$p^* = \begin{cases} 1 & \rho \le \sqrt{\frac{R}{T}} - 1 \\ \frac{\sqrt{\frac{R}{T}}-1}{\rho} & \rho > \sqrt{\frac{R}{T}} - 1 \end{cases}.$$

When $\rho \le \sqrt{\frac{R}{T}} - 1$, $p_e = p^* = 1$. Otherwise, $p_e > p^*$. It is possible to induce p^* in equilibrium by imposing an appropriate fee on trials or on service completions.

Sumita, Masuda and Yamakawa [166] considered a system with a finite buffer of size K. A customer who encounters a full buffer is rejected and obtains service from an alternative server. If the buffer is not full then the customer is accepted. To describe their model we define the following functions:

$V(\lambda)$ - aggregate utility gained by the arriving customers if all are accepted.
$R(\lambda)$ - aggregate utility gained by the arriving customers if all are rejected.
$\alpha(\lambda, \mu, K)$ - probability of acceptance.

$\beta(\lambda,\mu,K)$ - probability of rejection.
$G(\lambda,\mu,K)$ - expected cost incurred by accepted customer.
$M(\lambda,\mu,K)$ - expected cost incurred by rejected customer.
p_A - fee imposed on accepted customers.
p_R - fee imposed on rejected customers.

The social objective is to maximize $\alpha V + \beta R - \lambda(\alpha G + \beta M)$ subject to the equilibrium condition

$$\alpha V' + \beta R' = \alpha(p_A + G) + \beta(p_R + M).$$

An arriving customer causes negative externalities in two ways: by increasing waiting costs of other customers, and by increasing the probability of rejection for future customers. The authors call the latter type *loss externalities.* They proved that the equilibrium arrival rate, when no admission fees are imposed, is higher than the optimal arrival rate, and showed that this can be corrected by appropriate fees. They also solved the long-run problem in which both μ and K are decision variables and a cost $c(\mu, K)$ is added to the social objective function.

7. Multi-server models

7.1. Homogeneous service values

Bell and Stidham [27] analyzed an equilibrium model of routing customers in a queueing system. This section describes a simplified version of their model. Consider n exponential servers with service rates $\mu_1 \geq \mu_2 \geq \cdots \geq \mu_n$. For convenience, let $\mu_{n+1} = 0$. The arrival process is Poisson with rate Λ, and it is routed to the servers so that a customer is assigned to server i with probability $\frac{\lambda_i}{\Lambda}$, and consequently, the arrival process to server i $(i = 1,\ldots,n)$ is Poisson with rate λ_i. Balking is not allowed, and therefore $\sum_{i=1}^{n} \lambda_i = \Lambda$. We assume $\Lambda < \sum_{i=1}^{n} \mu_i$.

7.1.1 Equilibrium

In equilibrium, if $\lambda_i > 0$ then $\lambda_j > 0$ for $j = 1,\ldots,i$. Thus, in equilibrium, there exists an index i_e such that $\lambda_j > 0$ for $j \leq i_e$ and $\lambda_j = 0$ otherwise. Of course it is possible that all servers are active, i.e., $i_e = n$. Moreover, in equilibrium the active servers share the same expected waiting time, so that $\mu_j - \lambda_j = \mu_1 - \lambda_1$ for $j \leq i_e$, and $\mu_j < \mu_1 - \lambda_1$ for $j > i_e$. Apply $\sum_{j=1}^{i_e} \lambda_j = \Lambda$, to get

$$\lambda_i = \mu_i - \frac{\sum_{j=1}^{i_e} \mu_j - \Lambda}{i_e}, \qquad i = 1,\ldots,i_e. \tag{3.9}$$

Thus:

- In equilibrium, the excess rate of the active servers over total demand is equally distributed among the active servers.

By (3.9) and the fact that the expected waiting time at i_e is at least $\frac{1}{\mu_{i_e}}$, the expected waiting time at an active server i in equilibrium is

$$\frac{i_e}{\sum_{j=1}^{i_e}\mu_j - \Lambda} = \frac{1}{\mu_{i_e} - \lambda_{i_e}} > \frac{1}{\mu_{i_e}}.$$

On the other hand, if a customer joins an inactive server, then he only has to stay for his own service duration. Since no customer chooses to join an inactive server, and in particular server $i_e + 1$, it follows that

$$\frac{i_e}{\sum_{j=1}^{i_e}\mu_j - \Lambda} \le \frac{1}{\mu_{i_e+1}}.$$

These two inequalities imply that the equilibrium value of i_e is given by

$$i_e = \min\left\{ i \;\middle|\; \frac{1}{\mu_{i+1}} \ge \frac{i}{\sum_{j=1}^{i}\mu_j - \Lambda} \right\}. \tag{3.10}$$

7.1.2 Social optimization

The social problem is to set a routing vector $(\lambda_1, \ldots, \lambda_n)$ which solves the following problem:

$$\min \sum_{j=1}^{n} \frac{\lambda_j}{\mu_j - \lambda_j}$$

subject to

$$\sum_{j=1}^{n} \lambda_j = \Lambda,$$

and

$$0 \le \lambda_j < \mu_j, \quad j = 1, 2, \ldots, n.$$

Again, in an optimal solution a threshold index i^* exists, such that server j is active if and only if $j \le i^*$.

The contribution of an active server i to the social objective is $\frac{\lambda_i}{\mu_i - \lambda_i}$. In an optimal solution, the marginal values of these contributions, i.e., their derivatives as functions of the arrival rates, $\frac{\mu_i}{(\mu_i - \lambda_i)^2}$, $1 \le i \le i^*$, are all equal to some value, say α. Hence,

$$\lambda_i = \mu_i - \frac{\sqrt{\mu_i}}{\sqrt{\alpha}}, \quad i = 1, 2, \ldots, i^*. \tag{3.11}$$

Apply $\sum_{i=1}^{i^*} \lambda_i = \Lambda$ to get

$$\sqrt{\alpha} = \frac{\sum_{i=1}^{i^*} \sqrt{\mu_i}}{\sum_{i=1}^{i^*} \mu_i - \Lambda}$$

and hence

$$\lambda_i = \mu_i - \frac{\sqrt{\mu_i}}{\sum_{j=1}^{i^*} \sqrt{\mu_j}} \left(\sum_{j=1}^{i^*} \mu_j - \Lambda \right), \quad i = 1, \ldots, i^*.$$

- In a socially optimal solution, as in equilibrium, the input to an active server is equal to its service rate minus a portion of the excess of the total rate of service of the active servers over total demand. However, in contrast to the equilibrium solution, this excess is not distributed uniformly among the active servers when social optimization is sought, but in a way that is proportional to the square root of their service rates.

- The optimal value of i^* is obtained as for i_e in (3.10):

$$i^* = \min \left\{ i \;\middle|\; \frac{1}{\mu_{i+1}} \geq \frac{\sum_{j=1}^{i} \sqrt{\mu_j}}{(\sum_{j=1}^{i} \mu_j - \Lambda)\sqrt{\mu_i}} \right\}.$$

- By (3.11) the expected waiting time for a customer joining server i, $i \leq i^*$, is $\frac{\sqrt{\alpha}}{\sqrt{\mu_i}}$. Thus, under socially optimal routing, the waiting time at a slow server is greater than it is at a fast server.

- The optimal routing is not sustainable in equilibrium as customers will "migrate" from higher indexed servers to those with lower indices. It turns out that there exists an index i_t such that the arrival rates to servers with index at most i_t are larger in equilibrium than under social optimization, whereas the opposite holds with respect to the servers with an index greater than i_t. Thus, in equilibrium, customers overload the fast servers, relative to the socially optimal routing.

- $i_t \leq i_e \leq i^*$.

7.2. Heterogeneous service values

The model of Bradford [29] is similar to that of Section 3 in regard to the function $V'(\lambda)$. It assumes that there are m customer classes with Poisson arrivals, and class i has a service value function $V_i(\lambda_i)$. Service requirements are homogeneous. A central planner has to determine how much to charge a joining i-customer. The novelty of the model is the existence of n servers with different service rates. Let π_{ij} be the probability that an arriving i-customer is routed to queue j. Denote by $W_i(\underline{\lambda}, \pi)$ the expected waiting time of an i-customer, given an arrival rate vector $\underline{\lambda}$ and a routing matrix π. This value is a weighted average of waiting times at various queues.

The social optimization problem is

$$\max_{\underline{\lambda}, \pi} \sum_{i=1}^{m} [V_i(\lambda_i) - C_i \lambda_i W_i(\underline{\lambda}, \pi)]. \tag{3.12}$$

Assume now that the solution has positive arrival rates for each class (this holds when $V_i'(0)$, $1 \leq i \leq m$, are sufficiently large). The first-order conditions are

$$V_i'(\lambda_i) - C_i W_i(\underline{\lambda}, \pi) - \sum_{k=1}^{m} \frac{\partial W_k(\underline{\lambda}, \pi)}{\partial \lambda_i} C_k \lambda_k = 0, \quad 1 \leq i \leq m. \tag{3.13}$$

Let p_i denote the admission fee of an i-customer. Then, in equilibrium,

$$V_i'(\lambda_i) = C_i W_i(\underline{\lambda}, \pi) + p_i, \quad 1 \leq i \leq m.$$

Hence, prices which induce optimal social arrival rates satisfy

$$p_i = \sum_{k=1}^{m} \frac{\partial W_k(\underline{\lambda}, \pi)}{\partial \lambda_i} C_k \lambda_k, \quad 1 \leq i \leq m, \tag{3.14}$$

where the vector $\underline{\lambda}$ is determined by (3.13).

Suppose prices, routing probabilities, and arrival rates are given. A joining i-customer is asked for his class and is then charged and routed in accordance with the class he claims to belong to. Let $r_i(k)$ be the probability that an i-customer announces that he belongs to class k. Then the arrival rate to server-j is

$$\gamma_j = \sum_{i=1}^{m} \sum_{k=1}^{m} \lambda_i r_i(k) \pi_{kj}, \quad 1 \leq j \leq n. \tag{3.15}$$

Denote by $W_{(j)}(\gamma)$ the expected waiting time in queue-j, $j = 1, \ldots, n$, when the arrival rate to this queue is γ. Then,

$$W_i(\underline{\lambda}, \pi) = \sum_{k=1}^{m} r_i(k) \sum_{j=1}^{n} \pi_{kj} W_{(j)}(\gamma_j), \quad 1 \leq i \leq m.$$

For a given pricing and routing scheme, a set of arrival rates and reporting probabilities constitute an equilibrium if they solve

$$\min_{r_i(1), \ldots, r_i(m)} \sum_{k=1}^{m} r_i(k) \left[p_k + C_i \sum_{j=1}^{n} \pi_{k,j} W_{(j)}(\gamma_j) \right], \quad 1 \leq i \leq m, \tag{3.16}$$

subject to $\sum_{k=1}^{n} r_i(k) = 1$ for all $i = 1, \ldots, m$, and $r_i(k) \geq 0$ for all i and k, where γ_j is as in (3.15) and the arrival rates solve

$$V_i'(\lambda_i) = \sum_{k=1}^{m} r_i(k) \left[p_k + C_i \sum_{j=1}^{n} \pi_{k,j} W_{(j)}(\gamma_j) \right], \quad 1 \leq i \leq m.$$

Call a pricing and routing scheme *incentive-compatible* if $r_i(i) = 1$ (and hence $r_i(k) = 0$ for $k \neq i$) for all $1 \leq i \leq m$ minimizes (3.16). Bradford proved the following theorem.

THEOREM 3.6 *Let $\underline{\lambda}^*$ and π^* be the maximizers of (3.12). Then, the corresponding set of prices given in (3.14) are incentive-compatible in the sense that the optimal joining probabilities combined with customers' truthful revelation of their class type define an equilibrium.*

Consider now the problem faced by a facility manager trying to maximize his profit. The facility manager sets a menu of admission fees and routing probabilities. He is not able to distinguish customer classes but he knows the functions V_i. Each arriving customer then chooses from the menu, pays the fee and is routed to a server according to the probabilities corresponding to the stated class. Suppose that for $i = 1, \ldots, m$, i-customers reveal that they belong to class i, and choose (p_i, π_i) from the menu. In order for this to be an equilibrium the following conditions are necessary:

$$p_k - p_l \geq C_l[W_l(\underline{\lambda}, \pi) - W_k(\underline{\lambda}, \pi)], \quad k \neq l, \tag{3.17}$$

$$V_i'(\lambda_i) = p_i + C_i W_i(\underline{\lambda}, \pi). \tag{3.18}$$

Condition (3.17) means that a customer cannot increase his welfare by choosing a payment other than the one intended for his class. Condition (3.18) is the equilibrium condition for setting the arrival rates λ_i, $1 \leq i \leq m$.

By substituting (3.18) in (3.17) we obtain

$$V_k'(\lambda_k) - V_l'(\lambda_l) \geq (C_k - C_l) W_k(\underline{\lambda}, \pi), \quad k \neq l. \tag{3.19}$$

The objective of the queue manager is to maximize $\sum_i \lambda_i p_i$. Substituting (3.18) we get that the manager's optimization problem is

$$\max \sum_{i=1}^{m} [\lambda_i V_i'(\lambda_i) - \lambda_i C_i W_i(\underline{\lambda}, \pi)]$$

subject to (3.19).

Bradford proved that

- For every class, the arrival rate is no larger for the profit-maximizing solution than it is for the socially optimal one.

Suppose that $C_i < C_{i+1}$ $i = 1, \dots, m-1$, and $\lambda_i > 0$, $i = 1, \dots, m$. Bradford proved the following results:

- If (3.19) is satisfied for all (k, l) with $|k - l| = 1$ then it is also satisfied for every k, l.

- In a profit-maximizing solution, customers from class $i + 1$ are indifferent between choosing (p_i, π_i) and (p_{i+1}, π_{i+1}).

7.3. Class decision

Lee and Cohen [102] considered a set of n facilities, where the k-th facility is an $M/M/s_k$ queueing system with mean service time of $\frac{1}{\mu_k}$. The arrival process consists of m independent classes whose demands generate Poisson streams with rates γ_i, $i = 1, \dots, m$. The i-th stream is controlled by an *agent*. The agent selects rates x_{ik} and routes his customers so that the arrival process of i-customers to the k-th facility is Poisson with rate x_{ik}, $\sum_{k=1}^{n} x_{ik} = \gamma_i$. The aggregate arrival rate to facility k is $\lambda_k = \sum_{i=1}^{m} x_{ik}$.

Denote by $W_k(\lambda_k)$ the expected waiting time at the k-th queue. The expected waiting time for i-customers is then $\frac{1}{\gamma_i} \sum_{k=1}^{n} x_{ik} W_k(\lambda_k)$. The expected number of i-customers who are in service at any instance is $\sum_{k=1}^{n} \frac{x_{ik}}{\mu_k}$. Lee and Cohen assumed that agent i's objective is to set a routing vector x_{ik}, $k = 1, \dots, n$, so as to minimize a weighted sum of these two functions

$$\frac{b_{i1}}{\gamma_i} \sum_{k=1}^{n} x_{ik} W_k(\lambda_k) + b_{i2} \sum_{k=1}^{n} \frac{x_{ik}}{\mu_k},$$

given the routing vectors of the other agents. This is an m-person game. The main results are:

- There exists an equilibrium.

- If the facilities are identical ($s_k = s$ and $\mu_k = \mu$ for $k = 1, \dots, n$), then the equilibrium is unique.

- In general, the equilibrium allocation does not optimize social welfare, but in the case of identical facilities, it does.

8. Queueing networks

Queueing networks are a main area of research. Some work about optimal control of simple queueing networks has been done (see Stidham [164] for a survey) but only very little is known about equilibrium behavior in such models.

8.1. The Braess paradox

The work of Cohen and Kelly [36] extends the Braess paradox (§1.1) to queueing networks. The Braess paradox occurs in transport networks when the addition or expansion of a link generates a new equilibrium with higher travel costs to *all* users of the network. Cohen and Kelly demonstrated that a similar phenomenon is possible in queueing networks. We now outline a simplified version of their example.

Consider an arrival process of customers to a service system with an average rate of 1. The system consists of three facilities denoted by A, B and C. The service demand of a customer can be fulfilled by visiting C and either A or B. A and B are single server facilities with a service rate of 1, whereas C has an infinite number of servers. Let ω be the expected delay at A or B when the arrival rate to this facility is $\frac{1}{2}$. We assume that the service at C has a constant length of 2ω time units. In equilibrium the usage rates are $\frac{1}{2}$ both at A and B, and the expected total delay of each customer is 3ω.

Suppose now that a third option is added: service can also be fulfilled by visiting both A and B. The former solution no longer corresponds to an equilibrium since, when customers equally split between the former options, the new one guarantees a shorter delay of 2ω. In the new equilibrium the arrival rate of customers who select the new option is positive but less than 1 (had it been 1, the arrival and service rates at A and B would be equal, implying infinite delay). In the new equilibrium, the $A - B$ option is selected by a positive fraction of the customers, whereas the others equally split between the $A - C$ and $B - C$ options. Hence, the expected delay at A and B strictly exceeds ω. In equilibrium the expected delays are equal for all of the customers. In particular, those who select A and C incur a delay exceeding 3ω and therefore all customers suffer longer delays than in the original situation.

Calvert, Solomon, and Ziedins [32] show that the Braess paradox can also occur in observable queues. For further discussion on the Braess paradox see the survey by Altman, Boulogne, El Azouzi and Jimenez [8].

8.2. Heterogeneous service values

Masuda and Whang [120] generalized the model of Section 3 as follows. Let N be a finite set of identical servers. Customers are classified according to their needs. A k-customer can accomplish service by visiting a subset $R \in N$ from a given collection of subsets (called "routes" although there is no ordered "flow" in a "network" in this model) $\mathcal{R}_k$. Let $V_k(\lambda_k)$ be the aggregate service value to k-customers if their demand rate is λ_k. It is assumed that the expected waiting time at server i is a function $W_i(\lambda_i)$ of the total demand λ_i at the server. Selecting route $R \in \mathcal{R}_k$ is associated with a price $p(R)$. In equilibrium, for every route R used by k-customers,

$$V_k'(\lambda_k) = p(R) + C \sum_{j \in R} W_j(\lambda_j),$$

where C is the time cost. For routes not used by k-customers, the right-hand side is at least as large as the left-hand side. This equilibrium is also known as *Wardrop equilibrium* (see [172]).

Masuda and Whang proved the following results:

- Assume that the functions V_k' strictly decrease to 0, and the function W is strictly increasing. Then for any given set of route prices, $\{p(R)\}$, there exists a unique equilibrium.
- A socially optimal solution can be induced in equilibrium by setting route prices such that

$$p(R) = C \sum_{j \in R} \lambda_j^* W'(\lambda_j^*), \tag{3.20}$$

  where λ_j^* is the arrival rate to server j under a socially optimal solution. Thus, as in (3.4), the price is equal to a congestion externality cost caused by selecting route R.
- The route price is additive. Hence, it is sufficient to charge a service fee $C\lambda_j^* W'(\lambda_j^*)$ at server j for $j \in N$, where $p(R)$ is as in (3.20). Again this is an externality price. This is a more convenient charging method since the number of servers is typically much smaller than the number of routes.
- It is not always possible to induce social optimality by charging a fee that only depends on the type of customer (and not on his choice of route).

Masuda and Whang also investigated the long-run problem of setting optimal service rates (see §8), and extended some of the results discussed in §8.1. Note that because of the Braess paradox, for a fixed set of prices, it may be optimal to *reduce* the rates of some servers, even if this action does not yield direct savings in operation costs.

8.3. Serial networks with overtaking

Another model that deals with equilibrium in a queueing network is mentioned by Larson [100]. In this model, customers (vessels on inland U.S. waterways) move between queues (at locks) and select their travel speeds. High speed is associated with higher travel cost due to high fuel consumption. A customer who observes another customer just behind him increases his speed to avoid being overtaken and having to wait longer at the next queue. Larson claims that eventually customers move at their maximum speeds, and this is socially inefficient.[12] The model, however, is not formally described and analyzed, and it is not clear why travelling at maximum speed is indeed an equilibrium when there is some positive distance between customers (so that the one behind has no chance of overtaking the one in front). It may be that such an equilibrium is possible if customers' travel speeds differ or are subject to uncertainty. This is an interesting open problem.

9. Related literature

- Chen and Frank [33] discussed the robustness of the main result of Edelson and Hildebrand [47], that a profit maximizer chooses a socially optimal admission fee, when the assumption of a linear utility function is removed. A main issue here is how to model the social utility of the collected fees λp. Since these fees and the aggregate individual utilities are treated differently, there is no reason to expect that the socially optimal policy leaves customers with zero surplus. Therefore, the profit and welfare-maximizing admission fees differ. This is illustrated in [33] by an example that uses a specific utility function and the social value of the collected fees is assumed to be linear. Under certain conditions it is shown that a socially optimal fee may induce negative expected customer welfare. Such an outcome is natural when the customer's value of a monetary unit is smaller than its social value.

- Balachandran [16] considered an unobservable $M/G/1$ model with a fixed cost c of running the service facility. This cost does not depend on the number of customers served or their service times. Given an arrival rate λ, c is absorbed by the customers so that each individual

[12] This is an example of rent dissipation, see Remark 4.6.

pays $\frac{c}{\lambda}$. Balachandran investigated the impact of this cost allocation on the equilibrium arrival rate λ_e. An increase in λ affects the welfare of a customer in two ways: it increases his expected waiting costs, but it decreases his share in covering the operating cost c. Therefore, it is not clear a priori whether a joining customer gains or looses from an increase in λ.

Balachandran proved that the equilibrium arrival rate is unique, and then investigated the related question of how the equilibrium arrival rate behaves as a function $\lambda(c)$ of the operating cost. First, for any $c > 0$, $\lambda_e(c) < \lambda_e(0)$. This result is expected since with a positive operating cost each customer is worse-off in comparison to the zero operating cost case. Second, $\lambda_e(c)$ is monotone decreasing if and only if the total expected cost per unit of time due to queueing is greater than the expected cost of maintaining the service center while it is idle. In terms of λ_e this condition is

$$\lambda_e(c)CW_q(\lambda_e(c)) > c\left(1 - \frac{\lambda_e(c)}{\mu}\right).$$

Last, the class dominance property (see Section 4.1) does not hold in this model.

- Stidham [165] raised an issue that is commonly discussed in economic models. Consider a discrete time version of the model of Section 3. Fix a value for λ_t and suppose that the server sets a price p_t that maximizes his profits given λ_t. Suppose now that the arrival rate at period $t+1$ is set to the value λ_{t+1} that equates the marginal value of service with the full price, under the assumption that the arrival rate is λ_t, that is, $V'(\lambda_{t+1}) = p_t + W(\lambda_t)$. Suppose that this process repeats itself. Will p_t and λ_t converge to the optimal values? In particular, is this the case when the process initializes near the optimal values? In other words, is the optimal solution *stable*? Stidham showed that this may or may not be the case, and gave necessary and sufficient conditions for the $M/M/1$ model (see also Rump and Stidham [150]).

- Friedman and Landsberg [54] considered a discrete time model in which customers in period $n+1$ decide whether or not to enter the queue based on the expected delay, which they assume to be the same as the delay in the previous period n. They proved that if the capacity of the queue is sufficiently large, the equilibrium arrival rate is stable. For small capacities, the arrival rate typically oscillates near the equilibrium.

- The issue of stability of the equilibrium was further investigated by Masuda and Whang [120] under the assumption that the system manager does not have full knowledge of the demand. They considered alternative dynamic pricing rules and models of adaptive learning with bounded rationality, and characterized the equilibrium and its stability conditions.

Chapter 4

PRIORITIES

Priorities are often used in queueing systems as a mechanism for service allocation. By carefully selecting an appropriate system of priorities, both social welfare and profits can be increased.[1] This chapter also considers models where customers purchase priority. The rationale behind purchasing priority is twofold: priority enables overtaking ordinary customers who are present upon arrival, and it avoids being overtaken by later-to-arrive priority customers. The latter reason means that as more customers purchase priority, the more inclined an individual should be to do so himself in any given state. In other words, this is an FTC situation.

1. Observable queues

1.1. Equilibrium payments

The earliest work on equilibrium behavior in priority queues was published by Balachandran [15]. Consider an observable $M/M/1$ queue in which each arriving customer chooses from a discrete infinite set $\{b(0) < b(1) < \cdots\}$ of possible payments. Customers are assigned priority levels according to their payments and ties are broken in a FCFS manner. Balking is not allowed and customers are not informed on the payments made by others.[2]

---

[1] See the survey by Levhari and Sheshinski [105] for a discussion of these benefits.

[2] Balachandran claims that Theorem 4.1 also holds when an arriving customer knows the payments made by those already in the system. In particular, under the stated equilibrium, knowledge of n implies knowledge of the payments. This is correct, but care must be taken about predicting the behavior of future customers if the new arrival deviates from the equilibrium strategy.

The state of the system, $n \geq 0$, is defined in the case of a non-preemptive priority discipline as the number of customers in the queue, excluding the one in service. In the case of a preemptive priority discipline, this number includes the customer in service.

Consider the strategy of choosing the lowest payment that guarantees joining the head of the queue. If customers follow this strategy then a LCFS discipline is induced. Moreover, if this strategy is adopted and state n is observed then the payments made by the n customers are $b(0), \ldots, b(n-1)$, and in this order. Therefore, a payment of $b(n)$ guarantees that the arriving customer is positioned at the head of the queue. Theorem 4.1 gives sufficient conditions for this strategy to define an equilibrium:

THEOREM 4.1 *The strategy in which a customer who observes state n upon arrival chooses payment $b(n)$ is an equilibrium if and only if*

$$\max_{i\geq 1} [b(i) - b(i-1)] \leq \frac{C}{\mu(1-\rho)} \leq \frac{1+\rho}{\rho} \min_{i\geq 1} [b(i) - b(i-1)] .$$

Proof: We give the proof for the non-preemptive version (the proof for the preemptive version is similar). Tag a customer who chooses to deviate from the claimed strategy and offer a lower payment, say $b(n-r)$. He saves $b(n) - b(n-r)$ in direct payments but the time he spends in the system increases by r busy periods and hence his expected waiting costs increase by $\frac{Cr}{\mu(1-\rho)}$ (see (1.5)). Therefore, if the claimed strategy induces an equilibrium then the left inequality must hold.

Suppose now that the tagged customer deviates from the prescribed strategy and chooses to pay $b(n+1)$. If a new customer arrives before the service of the tagged customer is completed, then the new arrival will offer a payment of $b(n+1)$ and will not overtake the tagged customer. The expected reduction in waiting time is 0 if the service of the tagged customer ends before the next arrival, and it is the expected length of a busy period, $\frac{C}{\mu(1-\rho)}$, otherwise. The latter occurs with probability $\frac{\lambda}{\lambda+\mu} = \frac{\rho}{1+\rho}$. The condition for equilibrium is then

$$\frac{C\rho}{\mu(1-\rho)(1+\rho)} \leq b(n+1) - b(n).$$

It can be shown in a similar way that the same condition also guarantees that a a customer cannot profit by deviating to a payment $b(n+r)$ for $r > 1$.

□

The theorem was generalized to a $GI/G/1$ system by Balachandran and Srinidhi [24].

Note that since all customers are assumed to be identical, and balking is not allowed, there is no question regarding social optimality in this model.

Balachandran and Lukens [17] proved analogous conditions to those given in Theorem 4.1 for the following model, which is a mixture of the FCFS and LCFS disciplines. An arriving customer selects the same payment as the highest one paid by customers in the queue and joins the end of the segment of the queue of customers who paid this amount. If however, the size of this group exceeds a threshold, the arriving customer chooses a higher payment and then joins the head of the line, creating a new group.

Tilt and Balachandran [168] extended the model to allow customers with different time values. They showed that multiple pure equilibria may exist and that an equilibrium may induce various service orders, including FCFS and LCFS.

1.2. Two priority classes

Adiri and Yechiali [2] focused on the case of two priority classes. They assumed that the price for becoming a lower priority "ordinary customer" is 0. This assumption is without loss of generality since their model does not allow balking or reneging. The models described in the previous subsection assumed a solution, and gave conditions on the prices for priorities so that the solution defines an equilibrium. The approach of Adiri and Yechiali is different. They assumed a given price for priority and computed an equilibrium solution. In their model, two FCFS queues are formed in front of a single server, one line for priority customers and one for ordinary customers. The service of an ordinary customer is preempted if a priority customer arrives.[3] The preempted customer then moves to the head of the queue of ordinary customers and resumes service from the point where it was interrupted only when no priority customers are present. Upon arrival and after observing the length of the two queues, a customer decides whether to purchase priority. The price for priority is θ (measured in units of time). Customers cannot purchase priority while waiting.

The arrival process is Poisson with rate λ, the single server provides service whose length is exponentially distributed with mean $\frac{1}{\mu}$ regardless of priority level, and all customers bear a cost of C per unit of waiting

[3] Similar results hold in the model without preemption.

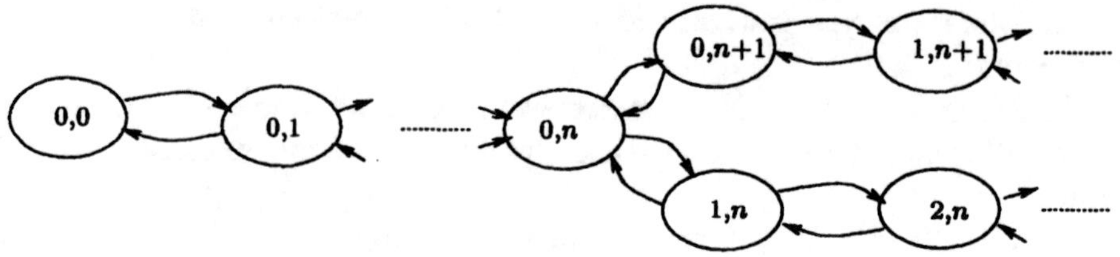

Figure 4.1. Transition diagram of a mixed threshold strategy

time. For stability, assume that $\rho = \frac{\lambda}{\mu} < 1$. No reneging or balking is allowed and therefore the value associated with service completion is immaterial.

Denote by (i,j), $i,j \geq 0$, a typical state of the system observed by an arrival, where i and j denote the number of priority and ordinary customers, respectively, in the system. A pure strategy prescribes for each state an action: to purchase priority or not. Evidently, the optimal action of a customer depends on the state of the system and the strategies adopted by future arrivals.

Adiri and Yechiali considered pure strategies of the following type: buy priority if and only if the number of customers in the queue is at least a threshold n.

If the system initializes with state $(0,0)$ and all follow the pure threshold strategy n, then the state space is basically one-dimensional since the only possible states are $(0,j)$, $j = 0,\ldots,n$ and (i,n), $i = 1,2,\ldots$, and from the total number of customers in the system one can infer the two-dimensional state. When the initial state in not $(0,0)$ (but still the threshold strategy n is used), then sooner or later, state $(0,0)$ will be reached and the above pattern will hold from then on. In other words, when all use the threshold strategy n, then the only recurrent states are $(0,j)$ $j = 0,\ldots,n$ and (i,n) $i = 1,2,\ldots$, whereas the other states are transient.

Hassin and Haviv [72] continued this line of research obtaining the results described in this section. They extended the set of possible strategies to include mixed strategies of the following kind: for some nonnegative real number $x = n+p$, where n is an integer and $0 \leq p < 1$, a customer who observes a total of k customers in the system joins the ordinary queue if $k \leq n-1$, does so with probability p if $k = n$, and otherwise buys priority. Note that when $p = 0$ we get the pure strategy with threshold n. Figure 4.1 depicts the transitions among states when $p > 0$. The strategy mixes between the two threshold strategies n (with probability $1-p$) and $n+1$ (with probability p). This can be seen in the figure, where the upper path corresponds to the pure strategy $n+1$ and the lower path corresponds to the pure strategy n.

For $n \geq 1$, let $W(n)$ be the expected time in the system of the last customer in the ordinary queue when the state is $(0, n)$ and all use the pure threshold strategy n. Let $B = \frac{1}{\mu-\lambda}$ be the expected length of a busy period (equivalently, the expected time it takes to reduce the number of customers in the system by one).

THEOREM 4.2 *The integer threshold strategy n, $n \geq 1$, specifies an equilibrium if and only if*

$$\theta + \frac{C}{\mu} - CB \leq CW(n) \leq \theta + \frac{C}{\mu}. \tag{4.1}$$

The threshold $n = 0$ specifies an equilibrium if and only if $\theta + \frac{C}{\mu} \leq CB$.

Proof: Assume that the entire population uses the integer threshold strategy n for some $n \geq 1$. For n to describe a best response strategy for an individual, two conditions are necessary. First, if he observes state $(0, n-1)$, not buying priority is optimal, that is,

$$CW(n) \leq \theta + \frac{C}{\mu}.$$

Second, if he encounters state $(0, n)$, his optimal action is to buy priority, that is

$$C[B + W(n)] \geq \theta + \frac{C}{\mu}.$$

Moreover, these conditions are also sufficient: if it is optimal to buy priority at $(0, n)$, it is also optimal to do so at (i, n) for $i \geq 1$; if it is optimal not to buy priority at $(0, n-1)$, it is also optimal not to do so at $(0, j)$ for $j \leq n-2$. To verify the first claim, note that purchasing priority in state (i, n) is associated with a cost of $\theta + \frac{C(i+1)}{\mu}$, whereas not buying priority costs $C[(i+1)B + W(n)]$. Since $B > \frac{1}{\mu}$, if the former is greater than the latter for $i = 0$, then this is clearly the case for $i > 0$. Similar arguments explain the second claim.

The fact that the threshold strategy $n = 0$ prescribes an equilibrium if and only if $\theta + \frac{C}{\mu} \leq CB$ is straightforward. □

THEOREM 4.3

- *For $n \geq 1$,*

$$\frac{1}{\mu} \leq W(n+1) - W(n) \leq B. \tag{4.2}$$

- $$\lim_{n\to\infty}[W(n+1) - W(n)] = \frac{1}{\mu}. \tag{4.3}$$

- *The number of pure threshold equilibria is between 1 and $\left\lfloor \frac{1}{1-\rho} \right\rfloor$.*

- *The lower and upper bounds of 1 and $\left\lfloor \frac{1}{1-\rho} \right\rfloor$ on the number of pure equilibria are attainable.*

Proof: The left inequality in (4.2) follows since the value of $W(n+1)$, in comparison to the value of $W(n)$, involves at least one more service completion. The right inequality can be argued as follows: $W(n) + B$ is the expected waiting time for an ordinary customer when n ordinary customers are ahead of him, no priority customers are present, and all use the threshold strategy n. $W(n+1)$ considers the same scenario but with one difference: all use the threshold strategy of $n+1$. This is of course more favorable for the ordinary customer in position $n+1$ and hence $W(n+1) \leq W(n) + B$.

The proof for (4.3) is technical. For details see [72]. Yet, this result has a simple explanation. Compare two ordinary customers, one in position n and one in position $n+1$ where in both cases no priority customers are present. The former customer is in a system where all use the threshold strategy n and the latter where all use $n+1$. Looking at the two under the same realization of events, the latter commences service for the first time as soon as the former leaves. When n is large it is unlikely that he will ever be preempted since the expected queue length behind him is bounded. Hence, his additional time in the system approaches $\frac{1}{\mu}$ when $n \to \infty$.

The third claim of the theorem follows from (4.2) and the fact that the interval bounding $W(n)$ for an equilibrium n (as defined by (4.1)) is of width B. Thus, the number of pure equilibria is at least one and at most $\left\lfloor \frac{B}{1/\mu} \right\rfloor$ which equals $\left\lfloor \frac{1}{1-\rho} \right\rfloor$.

For the fourth claim, when $\theta = 0$ there is exactly one equilibrium (at $n = 0$), leading to the stated lower bound. The upper bound is attained

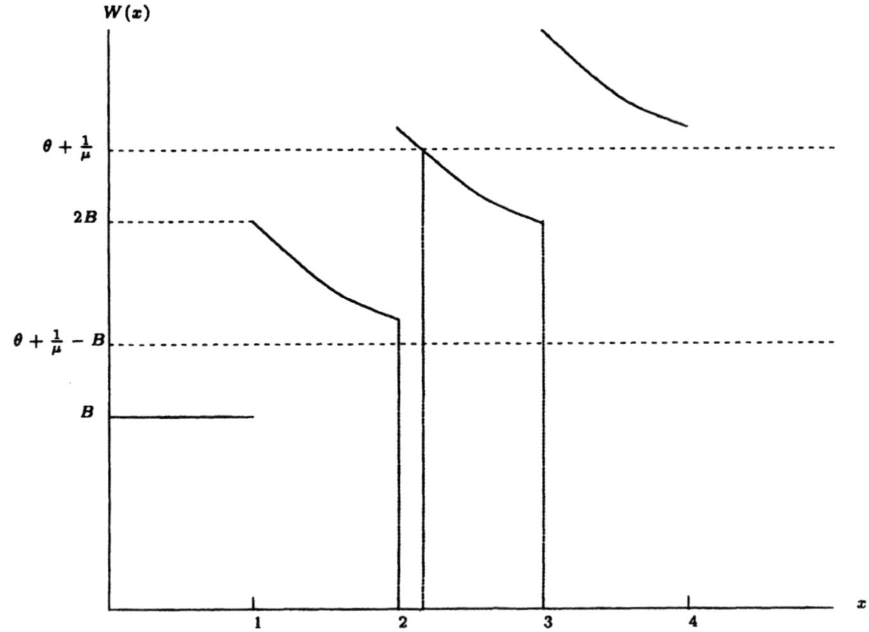

Figure 4.2. Pure and mixed equilibria

when $\theta \to \infty$. This is a consequence of the limit stated in (4.3) and the fact that the width of the interval determined by (4.1) equals $\frac{1}{\mu(1-\rho)}$. □

For a non-integer value $x > 0$, let $W(x)$ be the expected time in the system for the last customer in the ordinary queue when the state is $(0, \lceil x \rceil)$ and all use the threshold strategy x. A necessary condition for x to define an equilibrium strategy is that a customer who observes state $(0, \lfloor x \rfloor)$ upon arrival is indifferent between buying priority and not (and therefore randomizing between the two options is a best response). In fact, this condition is sufficient since it also implies that each of the other actions prescribed by the strategy for any recurrent state is a best response. The next theorem follows from this observation. The left-hand side of the expression is the expected cost associated with not buying priority and the right-hand side gives the expected cost associated with buying it.

THEOREM 4.4 *The non-integer threshold x defines an equilibrium strategy if and only if*

$$CW(x) = \theta + \frac{C}{\mu}.$$

Figure 4.2 illustrates the function $W(x)$. The equilibria in this example correspond to pure strategies $x = 2$, $x = 3$ and a mixed one where $2 < x < 3$. Several observations follow:

- For $0 < x \leq 1$, $W(x) = B$. The customer in question is the first in the ordinary queue, and until his departure every new arrival will join the priority queue and overtake him. Thus, his waiting time equals a busy period.

- For a fixed value of $\lceil x \rceil$, $W(x)$ is continuous and monotone decreasing in x. This holds because as x increases, customers are less likely to purchase priority, thereby reducing the expected future waiting time of the customer currently in position $\lceil x \rceil$ of the ordinary queue.

- $W(n+) - W(n) = B$. $W(n+)$ equals the expected length of a busy period B, which is the time it takes a customer in position $n+1$ to reach position n when (almost) all use strategy n, plus the expected waiting time from position n under (basically) the same conditions, i.e., $W(n)$.

- $W(n+) - W(n+1) \leq B - \frac{1}{\mu}$. This holds since

  $$W(n+) - W(n+1) \leq W(n+) - W(n) - \frac{1}{\mu} = B - \frac{1}{\mu}.$$

  The inequality here follows from Theorem 4.3 and the equality follows from the previous observation.

- The sequence of equilibrium strategies alternates between pure and mixed strategies. The exception is when $W(n+) = \theta + \frac{1}{\mu}$ for some integer n (which is atypical and the data have to be specifically selected in order for this phenomenon to occur). In such cases consecutive pure equilibrium thresholds exist.

- As observed, $W(n+) - W(n) = B$, whereas the decrease in the function between consecutive integers satisfies $W(n+) - W(n+1) < B$. Therefore, if the equilibrium condition for a mixed strategy at x is satisfied, then both $\lfloor x \rfloor$ and $\lceil x \rceil$ are pure equilibria. In particular, both the smallest and the largest x values that correspond to equilibrium strategies are integers. Moreover, *the set of thresholds that define pure equilibria consists of consecutive integers.*

$W(x)$ is computed as follows. Let $H^x_{i,k}$ be the (future) expected waiting time for an ordinary customer at position i in his queue ($i = 1$ corresponds to a customer in service), when k ordinary customers are behind him, the priority queue is empty, and the threshold x is used. Of course, $i \geq 1$, $k \geq 0$ and $i + k \leq \lceil x \rceil$. The H values can be computed by solving a set of linear equations, as shown below. Given these values, $W(x)$ is computed by setting $W(x) = H^x_{\lceil x \rceil,0}$.

Assume that a threshold $x = n + p$, $0 < p \leq 1$ is adopted by the population. Then the maximum possible length of the ordinary queue (including the customer in service) is $\lceil x \rceil = n+1$, leading to the following set of equations:[4]

$$\begin{aligned}
H^x_{1,n} &= B, \\
H^x_{1,n-1} &= 1 + \lambda p H^x_{1,n} + \lambda(1-p)(B + H^x_{1,n-1}), \\
H^x_{1,k} &= 1 + \lambda H^x_{1,k+1} \quad k = 0, 1, \ldots, n-2, \\
H^x_{i,k} &= 1 + \lambda H^x_{i,k+1} + \mu H^x_{i-1,k} \quad i = 2, \ldots, n-1 \quad k = 0, \ldots, n-i-1, \\
H^x_{i,n-i} &= 1 + \lambda p H^x_{i,n-i+1} + \lambda(1-p)(B + H^x_{i,n-i}) + \mu H^x_{i-1,n-i} \quad i = 2, \ldots, n, \\
H^x_{i,n-i+1} &= 1 + \lambda(B + H^x_{i,n-i+1}) + \mu H^x_{i-1,n-i+1} \quad i = 2, \ldots, n+1.
\end{aligned}$$

An $O(n^2)$ recursive solution procedure that takes advantage of the special structure of these equations is suggested in [72]. In particular, the system possesses a unique solution.

We conclude this section with several observations.

- Let n be such that both inequalities in (4.1) are strict. Then among the threshold strategies, n is the unique best response for an individual against all others using the threshold strategy n. The situation is different when the equilibrium is based on a non-integer threshold x: if all customers deviate to any threshold strategy $x - \eta$ (respectively, $x + \eta$) for any $\eta > 0$ then $\lfloor x \rfloor$ (respectively, $\lceil x \rceil$) is now a strictly better response against x than x itself. Thus, when limiting ourselves to threshold strategies, pure equilibria are ESS whereas mixed equilibria are not.

- The threshold strategy n means that customers purchase priority if they observe n or more customers in the system. Suppose that this is an equilibrium strategy. Consider now the following non-threshold strategy: purchase priority if the number of customers is exactly n. This too is an equilibrium. This may seem strange since it prescribes not to purchase priority when the number of customers is large. However, assuming the system initializes in state $(0,0)$, then if all follow this non-threshold strategy, states of the type $(i, n+j)$ with $i \geq 0$

[4]To simplify the presentation, we select the time units so that $\lambda + \mu = 1$. In the case of a pure strategy, with $p = 0$, the first and the last equations are not relevant.

and $j \geq 1$ will never be reached and hence it does not matter what is prescribed there as long as the strategy agrees with the threshold strategy for states $(0, i)$, $0 \leq i \leq n$. In particular, the two strategies lead to the same progression of events. If the initial state is not $(0, 0)$, the above is still true in the long-run since the states $(i, n + j)$ with $i \geq 0$ and $j \geq 1$ are transient under the two strategies.

- The previous item described a non-threshold equilibrium strategy which is not subgame perfect (SPE). We observe that the threshold equilibrium strategies are not necessarily SPE. The threshold strategy considers only the number of customers in the system, and not their type. It is easy to construct examples where the type matters. For example, consider a case where $\lambda \approx 0$. Then, if only priority customers are present then clearly there is no gain associated with buying priority. This is, of course, not the case if the present customers are ordinary.
- Suppose that the value of service is R, and that customers can balk. Clearly, the priority queue will never exceed $m = \left\lfloor \frac{(R-\theta)\mu}{C} \right\rfloor$. If $m \geq 1$ then an arrival who observes m priority customers in the queue balks. If $m = 0$ then customers never buy priority, and they join the ordinary queue as long as its length is at most $\left\lfloor \frac{R\mu}{C} \right\rfloor - 1$.
- Let $B_m = \frac{1}{\mu} \sum_{i=0}^{m-1} \rho^i$ be the expected length of a busy period in the $M/M/1/m$ queue (that is, an $M/M/1$ queue with a buffer of length m, including the customer in service). The analysis for the model that allows balking is similar to the one which forbids balking. In particular, this analysis can be achieved by replacing B with B_m.
- Hassin and Haviv designed an example with an equilibrium strategy in which the set of recurrent states is different from the one induced by any threshold strategy. The example assumes that balking is allowed.

1.3. Profit maximization

Alperstein [7] considered the model of the previous subsection, but with several priority levels managed by a profit maximizing server. Alperstein proved that the profit maximizing pricing scheme is such that the resulting equilibrium is with a threshold of 1 for each priority type except for the highest one. When the number of priority types is unbounded, this means that the service discipline generated in equilibrium is LCFS-PR. Alperstein also proved that the profit increases with the number of priority levels and that for an unbounded number of such levels, customers' surplus is 0.

These results are related to Hassin's claim that in a LCFS-PR exponential queue, customers' behavior is socially optimal (see §2.2). However, as in the other observable models discussed in this section, Alperstein's model does not allow balking or reneging so the question of socially optimal control of the queue is not relevant.

2. Unobservable queues

We now consider the unobservable version of the model discussed in Section 1.2, where customers decide whether to purchase priority at the price of θ without first observing the queue. A strategy is associated with a probability p of purchasing priority.

There are two ways in which an increase in the parameter p, used by all other customers, affects a given customer. On the one hand, it increases the expected number of customers who overtake an ordinary customer and hence increases the incentive to purchase priority. On the other hand, it decreases the expected number of ordinary customers that a priority customer expects to overtake and hence decreases the incentive to purchase priority. The latter effect does not exist in the observable model since there, the number of ordinary customers to be overtaken is known at the time of making the decision. Therefore, unlike the observable model, it is not intuitively clear here if the model is of the FTC type. It turns out however, that it is, as proved in the next theorem.

Under the FTC assumption, there are three possibilities for an equilibrium, exactly as in the shuttle model of §1.5: there may be a unique equilibrium with either $p = 0$ or $p = 1$, or there may be three equilibria, $p = 0$, $p = 1$ and a third p_e such that $0 < p_e < 1$. The precise conditions are listed in the following theorem:

THEOREM 4.5

- *The model is of the FTC type.*
- *If* $\theta \leq \frac{C\rho}{\mu(1-\rho)}$ *then* $p = 1$ *is a dominant strategy;*
- *If* $\theta \geq \frac{C\rho}{\mu(1-\rho)^2}$ *then* $p = 0$ *is a dominant strategy;*
- *If* $\frac{C\rho}{\mu(1-\rho)} < \theta < \frac{C\rho}{\mu(1-\rho)^2}$ *then there are three equilibria:* $p = 0$, $p = 1$ *and* $p = \frac{1}{\rho} - \frac{C}{\theta\mu(1-\rho)}$.
- *The pure equilibria are ESS. Mixed equilibria are not.*

Proof: Given that strategy p is adopted, the arrival process of priority customers is Poisson with rate λp and therefore the expected waiting time of a priority customer is $\frac{1}{\mu-\lambda p}$. Let W denote the expected waiting time of an ordinary customer. Then, the expected waiting time of a random customer can be computed in two ways. First, it equals the expected waiting time in a FCFS queue, which is $\frac{1}{\mu-\lambda}$. Second, it is $\frac{1}{\mu-\lambda p}$ with probability p and W with probability $1-p$. Thus,

$$\frac{1}{\mu-\lambda}=\frac{p}{\mu-\lambda p}+(1-p)W,$$

from which we conclude that $W=\frac{\mu}{(\mu-\lambda)(\mu-\lambda p)}$.[5]

The reduction, $f(p)$, in expected waiting costs due to becoming a priority customer is

$$f(p)=\frac{\lambda C}{(\mu-\lambda)(\mu-\lambda p)}=\frac{\rho C}{\mu(1-\rho)(1-\lambda p)}.$$

This function is monotone increasing with p, from which the FTC property follows.

Hence,

- When $\theta \leq f(0)$, it is uniquely optimal for a customer to purchase priority no matter what the others do. In other words, this is a dominant strategy.
- When $\theta \geq f(1)$, it is uniquely optimal for a customer not to purchase priority no matter what the others do. Again, this is a dominant strategy.
- When $f(0)<\theta<f(1)$ there exists a unique value p_e, $0<p_e<1$, such that given that p_e is adopted by the other customers, a customer is indifferent between purchasing priority or not. Solving $f(p)=\theta$ leads to $p_e=\frac{1}{\rho}-\frac{C}{\theta\mu(1-\rho)}$. This equilibrium is not ESS: if p is larger than p_e, the unique best response is to buy priority, whereas if p is smaller than p_e, then the unique best response is not to buy. Therefore, in addition to the mixed equilibrium, there are two pure equilibria, where all buy priority and where none do. These pure equilibria are ESS.

□

Figure 4.3 depicts the best response function for the three cases.

REMARK 4.6 The second item of Theorem 4.5 means that if θ is smaller than the expected queueing cost under the FCFS discipline, then all

[5] For an alternative derivation see, for example, [89] p. 125.

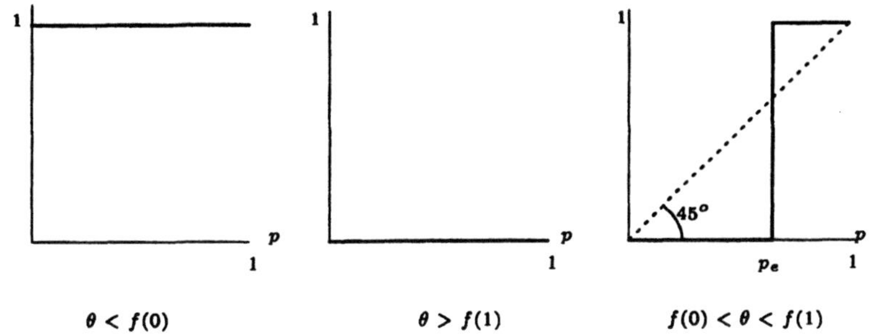

Figure 4.3. Best response vs. fraction of priority customers

customers will purchase priority in equilibrium. Thus, the additional option of buying priority makes everybody worse-off: all customers pay θ but in practice nobody gains from it. This is an example of *rent dissipation*, see [94, 169]. Similarly, when $f(0) < \theta < f(1)$, a fraction of the customers purchases priority in equilibrium; those who do not purchase priority are worse-off because they are pushed to the back of the queue; those who purchase priority have in equilibrium the same expected net benefit as those who do not, and therefore these customers are also worse-off.

We have seen that in both the observed and the unobserved models the situation is of the FTC type. Agastya [4] suggested a static model which yields a different outcome. Suppose that $n+1$ customers are present in a queue. Tag one of the customers and assume that of the other n customers x are ordinary customers and $n-x$ are priority customers. The service order in each class is random. If the tagged customer buys priority then his expected queueing time is $\frac{x}{2\mu}$. Otherwise, if he joins as an ordinary customer, his expected queueing time is $(x+\frac{n-x}{2})\frac{1}{\mu}$. The expected amount saved when buying priority is equal to $\frac{n}{2\mu}$. An interesting outcome is that this saving is independent of x!

Suppose now that each of the $n+1$ customers possesses the option of paying θ for priority. Then, if $\theta < C\frac{n}{2\mu}$ (respectively, $\theta > C\frac{n}{2\mu}$) there is a unique equilibrium in which all (respectively, none) buy priority. If $\theta = C\frac{n}{2\mu}$ then any strategy is an equilibrium.

3. Discriminatory processor sharing

Processor sharing relates to situations in which servers split their capacity among the customers who are present in the system. Thus, the larger the number of customers in the system, the slower the service rate they receive. In §1.4 and in §3.5 we referred to an *egalitarian processor sharing* model (EPS) in which customers evenly share the capacity of a single server. A more general model is the *discriminatory processor sharing* model (DPS) in which each customer owns a *relative priority* parameter and receives service proportionally to his parameter. Thus, if n customers are present and their priority parameters are $x_1, x_2, \ldots, x_n$, with $x_i \geq 0$ for $1 \leq i \leq n$, then customer i receives service at a rate of $\frac{x_i}{\sum_{j=1}^{n} x_j}$ of the server's capacity.[6] Of course, EPS is a special case of DPS in which the relative priorities are identical.

We next deal with two unobservable DPS models. In the first, only two parameters are available whereas in the second, customers are free to choose any nonnegative value. In both models the arrival process is Poisson with parameter λ, service requirements are exponentially distributed with parameter $\mu > \lambda$ (that is, $\rho < 1$), the common time value is C per time unit, balking and reneging are not permitted, and customers are not aware of their own service requirement.

3.1. Two relative priority parameters

Let $x_1 > x_2 \geq 0$ be the two relative priority parameters available. Without loss of generality assume that $x_1 + x_2 = 1$. Each customer has a choice of paying an amount of $\theta > 0$ and obtaining the priority parameter x_1, or else getting the priority parameter x_2. This is an extension of the model considered in the previous section, where $x_1 = 1$ and $x_2 = 0$, since within classes the order of service is FCFS.

A strategy is characterized by the probability q of purchasing x_1. Based on Fayolle, Mitrani and Iasnogorodski [51] the expected waiting time for an x_1-customer is

$$\frac{1}{\mu - \lambda}\left[1 - \frac{\lambda(1-q)(x_1 - x_2)}{\mu - \lambda(x_1 q + x_2(1-q))}\right], \tag{4.4}$$

and for an x_2-customer it is

$$\frac{1}{\mu - \lambda}\left[1 + \frac{\lambda q(x_1 - x_2)}{\mu - \lambda(x_1 q + x_2(1-q))}\right]. \tag{4.5}$$

Let $f(q)$ be the reduction in the expected waiting costs for an x_1-customer in comparison with an x_2-customer, when all use the strategy

[6] A customer with $x_i = 0$ receives service only when no customers with positive parameters are present. If all present customers have $x_i = 0$ then they evenly share the capacity of the server.

q. Then from (4.4) and (4.5),

$$f(q) = \frac{C}{\mu - \lambda} \frac{\lambda(x_1 - x_2)}{\mu - \lambda(x_1 q + x_2(1-q))}.$$

Since $f(q)$ is monotone increasing in q, $0 \le q \le 1$, the model is of the FTC type. The next theorem generalizes Theorem 4.5. Note that Remark 4.6 also holds for this model.

THEOREM 4.7

- *If $\theta < f(0)$, $q = 1$ is a dominant strategy.*
- *If $\theta > f(1)$, $q = 0$ is a dominant strategy.*
- *If $f(0) < \theta < f(1)$, there are three equilibria: $q = 0$, $q = 1$ and q_e where $f(q_e) = \theta$. Specifically,*

$$q_e = \frac{1}{\rho(x_1 - x_2)} - \frac{C}{(\mu - \lambda)\theta} - \frac{x_2}{x_1 - x_2}.$$

- *The pure equilibria are ESS. The mixed equilibrium is not.*

3.2. A continuum of relative priority parameters

Haviv and van der Wal [79] allowed customers to purchase relative priorities for any nonnegative amount of their choice, and the relative priority obtained for a price $p \ge 0$ is a monotone increasing function $x(p)$, with $x(0) = 0$. The inverse function $x^{-1}(y)$ denotes the price one pays to receive priority y.[7]

3.2.1 A linear price function

We start with the simpler case where $x(p) = p$.

THEOREM 4.8 *For a DPS system with $x(p) = p$, there is a unique equilibrium in which all customers pay*

$$\frac{\rho C}{\mu(1-\rho)(2-\rho)}.$$

The proof starts with several observations and lemmas.

[7] Haviv and van der Wal also treated a variation of this model in which each customer receives service individually but the order of entrance to service is determined probabilistically by relative priorities.

Let $W(n,p)$ be the expected residual waiting time of a customer who paid p, given that all others pay 1 and that he is in service with n other customers. Observe that $W(n,0)$ is the expected sum of $n+1$ busy periods and thus

$$W(n,0) = \frac{n+1}{\mu - \lambda}.$$

When $p \to \infty$ this customer gets absolute priority, and therefore $W(n,\infty) = \frac{1}{\mu}$. Thus, for $0 < p < \infty$

$$\frac{1}{\mu} < W(n,p) < \frac{n+1}{\mu - \lambda}.$$

For $p > 0$ and $n \geq 0$, $W(n,p)$ satisfies the difference equation:[8]

$$(\lambda + \mu)W(n,p) = 1 + \lambda W(n+1,p) + \mu \frac{n}{n+p} W(n-1,p). \qquad (4.6)$$

LEMMA 4.9 *For some functions $A(p)$ and $B(p)$, and every $n \geq 0$,*

$$W(n,p) = A(p)n + B(p). \qquad (4.7)$$

Proof: At a given time, say time 0, tag a customer who paid p, when there are n other customers in the system and all other customers (including future arrivals) pay 1. Suppose that an additional customer joins the system. The lemma claims that the added expected waiting times inflicted on the tagged customer due to the additional customer is independent of n. To prove this, suppose that service is given in a weighted round-robin fashion, namely for some (small) quantum Δ, the server, in a cyclical order, dedicates the service capacity for an amount of time Δ to each of the other customers and then serves the tagged customer for an amount of time $p\Delta$. Consider the service times of the additional customer, those who arrive during his service, those who arrive during their periods of service,... , ad infinitum. The tagged customer may be present during some of these periods but the distribution of his added time in the system due to the extra customer is independent of how many others are present in the system at time 0. Finally, look at the expected waiting time inflicted by the added customer when Δ goes to 0. This is exactly the value of $A(p)$. □

[8]The coefficient of $W(-1,p)$ is 0 so there is no need to define this term.

LEMMA 4.10 *The functions $A(p)$ and $B(p)$ defined in Lemma 4.9 are*

$$A(p) = \frac{1}{\mu(1+p-\rho)}, \tag{4.8}$$

and

$$B(p) = \frac{1+p}{\mu(1+p-\rho)}. \tag{4.9}$$

Proof: (4.8) and (4.9) follow by substituting (4.7) into (4.6). In particular, select any pair of values for n, such as 0 and 1, and get two linear equations for $A(p)$ and $B(p)$, which are solved by (4.8) and (4.9). □

LEMMA 4.11 *Denote by $g(p)$ the expected time in the system for a customer who pays p when the other customers pay 1. Then,*

$$g(p) = \frac{1+p-\rho p}{(1+p-\rho)(1-\rho)\mu}. \tag{4.10}$$

Proof: The distribution of the number of customers in the system is as in an $M/M/1$ model with the same arrival rate and mean service time. Therefore, the probability that a customer observes $n \geq 0$ other customers in the system upon his arrival is $(1-\rho)\rho^n$, see (1.2). If he pays p then his expected waiting time is

$$g(p) = A(p)\sum_{n=0}^{\infty}(1-\rho)\rho^n n + B(p)$$

which along with (4.8) and (4.9) gives (4.10). □

REMARK 4.12 Note that $g(0) = \frac{1}{\mu(1-\rho)^2}$ is equal to the expected time in an $M/M/1$ system from the arrival of a customer until the server is idle for the first time, see (1.6); $g(1) = \frac{1}{\mu(1-\rho)}$ is the expected waiting time in an $M/M/1$ EPS, see (1.4); when $p \to \infty$, $g(p) \to \frac{1}{\mu}$.

Proof of Theorem 4.8: Suppose that all customers pay the price of 1. In order for 1 to be the optimal price for a tagged customer, it is necessary that

$$\frac{d}{dp}(Cg(p)+p)\Big|_{p=1} = 0, \tag{4.11}$$

where $g(p)$ is given in (4.10). Since $g(p)$ is convex, this condition is also sufficient. Straightforward differentiation shows that if

$$C = \frac{\mu(1-\rho)(2-\rho)}{\rho},$$

then (4.11) holds. A change of scale concludes the proof. □

3.2.2 A concave price function

Now we are ready to generalize the result to any strictly monotone increasing and concave function $x(p)$.

THEOREM 4.13 *If $x(p)$ is concave then there exists a unique equilibrium priority level x_e, where x_e is the unique solution to*

$$x_e \frac{d}{dy} x^{-1}(y)\Big|_{y=x_e} = \frac{C\rho}{\mu(1-\rho)(2-\rho)}.$$

Proof: Let $\hat{W}(b,a)$ be the expected time in the system for a customer who buys priority level b when everybody else buys priority level a. Note that $\hat{W}(b,a) = g\left(\frac{b}{a}\right)$ where g is defined in Lemma 4.11. Replacing p in (4.10) with $\frac{b}{a}$, and taking the derivative with respect to b we get

$$\frac{\partial}{\partial b}\hat{W}(b,a) = -\frac{\frac{\rho}{a}(2-\rho)}{\left(1-\rho+\frac{b}{a}\right)^2 \mu(1-\rho)}.$$

The cost for a customer whose priority is b when all others have priority a is $C\hat{W}(b,a) + x^{-1}(b)$. By Lemma 4.11 and the concavity of $x(p)$, this function is convex in b (for any given a). Differentiating with respect to b and equating to 0 gives

$$C\frac{\frac{\rho}{a}(\rho-2)}{\left(1-\rho+\frac{b}{a}\right)^2 \mu(1-\rho)} + \frac{d}{db}x^{-1}(b) = 0. \tag{4.12}$$

When $b = a$ we get that in equilibrium

$$a\frac{d}{da}x^{-1}(a) = \frac{\rho C}{\mu(1-\rho)(2-\rho)}.$$

□

For example, if $x(p) = p^{\frac{1}{n}}$, then

$$x_e = \frac{\rho C}{n\mu(1-\rho)(2-\rho)}.$$

It is not intuitively clear whether this model is ATC, FTC, or neither of these. Consider the implicit function of b in the variable a, which is defined through (4.12):

$$\frac{d}{db}x^{-1}(b)\left[\frac{b^2}{a} + (1-\rho)^2 a + 2(1-\rho)b\right] = \frac{(2-\rho)\rho}{\mu(1-\rho)}C. \tag{4.13}$$

The term in the square brackets increases with b. For any given value of b, this term decreases with a as long as $0 < a < \frac{b}{1-\rho}$, and in particular, when $a = b$. It is monotone increasing for $a > \frac{b}{1-\rho}$. This, coupled with the convexity of the inverse function x^{-1}, implies that in order to maintain equality in (4.12) and in (4.13), one needs to increase b if $0 < a < \frac{b}{1-\rho}$, and decrease it if $a > \frac{b}{1-\rho}$. In equilibrium $a = b$ so that around this point there is an FTC situation.

4. Incentive compatible prices

The design of optimal price mechanisms may require knowledge of characteristics of the customers. These characteristics are seldom known to the queue manager and must be estimated or obtained from the customers' own statements, a situation that may create an incentive for customers to declare untruthful values. The topic of this section is the design of *incentive compatible prices* that not only induce optimal behavior from the central planner's point of view when the true parameter values are given to him, but also induce truthful statements from the customers as part of the resulting equilibrium.

4.1. Heterogeneous time values

The simplest and most intuitive model of an incentive compatible pricing scheme is that of Ghanem [55]. The model assumes an unobservable $M/G/1$ facility where the time value of customers is a continuous nonnegative random variable whose value is known to the customer but not to the system's manager. Balking is not allowed and the objective of the system's manager is to minimize the expected waiting costs by assigning customers to m priority classes and using a non-preemptive priority discipline. The parameter m is exogenously given.

Ghanem first proved the intuitive result that for social optimization higher priority should be given to customers with a higher time value (this is the $C\mu$-rule). Therefore, an optimal solution is characterized by numbers $\alpha_1 > \alpha_2 > \cdots > \alpha_{m-1}$ where α_i is the separation point between priority i and $i+1$, and customers with $\alpha_i \leq C < \alpha_{i-1}$, $1 \leq i \leq m$, (with $\alpha_0 = \infty$ and $\alpha_m = 0$) should obtain priority i, which is higher than priority $i+1$.

To induce customers to state their true classes, admission fees $p_1 > p_2 > \cdots > p_m$ are announced. Let $W_1 < W_2 < \cdots < W_m$ be the expected waiting times at each priority queue. These values are computed under the assumption that customers join their classes according to the socially optimal solution.

Since α_i is the optimal separation point between queue i and queue $i+1$, a customer with $C = \alpha_i$ is indifferent between joining queue i and joining queue $i+1$. Thus, $p_i + \alpha_i W_i = p_{i+1} + \alpha_i W_{i+1}$, or

$$p_i - p_{i+1} = \alpha_i(W_{i+1} - W_i), \quad i = 1, \ldots, m-1.$$

Note that one of the fees can be set in an arbitrary way, say $p_m = 0$, since balking is not permitted and only the differences in fees matter. It is straightforward that under this pricing scheme, customers with $C < \alpha_i$ prefer priority $i+1$ to i and those with $C > \alpha_i$ have the opposite preference. Hence, given these prices, individual optimization minimizes the systemwide delay costs.[9]

4.2. Pricing based on externalities

Dolan [45] also considered a model in which customers are identical except for having different time values, and priority levels are assigned to customers according to their declared value of time. The information that is available to an arriving customer includes the queue length upon his arrival, the declared time values of the customers in the queue, and the residual service time of the customer in service. Using this information, the customer declares a value to his time and obtains priority accordingly. The goal is to induce customers to declare their true time value so that the resulting order of service optimizes social welfare. Dolan suggested the use of *Clarke prices*. The idea is that each customer pays for the costs he imposes on others (both past and future arrivals) when joining the queue. These costs are calculated based on the assumption that customers reveal their true time values. If this assumption is correct then the cost paid by a customer is equal to the externalities he imposes. It follows that under this pricing rule, the strategy that prescribes declaring the true time value defines an equilibrium.

To explain this result, consider first a static model in which no new arrivals are expected. Suppose that all customers declare their true values, $C_1 > \cdots > C_k$. If customer i increases his declaration to a value C such that $C > C_{i-1}$ then he overtakes customer $i-1$. This decreases his waiting time by say, τ time units, at the expense of a similar increase in customer $i-1$'s waiting time. He directly saves $C_i\tau$ but there will be an increase in the price he pays by an amount of $C_{i-1}\tau > C_i\tau$, which is the cost he imposes on customer $i-1$. Hence his net gain is negative. Similarly, a decrease in the stated value below C_i adds waiting costs which are larger than the decrease in his payment.

[9] Ghanem derived explicit solutions for $m = 2$ and uniform or exponential time value distributions.

Generally, if a customer overstates his time cost he must compensate the customers who will have to wait longer as a result of this act. However, the total waiting time of customers in the system is independent of the declaration, so that the extra wait of these customers is exactly what the current one saves. Since these customers have higher time values than his, compensating for their increased waiting time costs the customer more than what he saves. Similarly, understating the time value adds waiting costs to the customer while reducing his payment. Yet, the saved payments were used to pay for the time of customers with lower time values so that again the net gain is negative.

This argument is general and holds for both absolute and relative priority regimes, and for observable as well as unobservable models.

4.3. Heterogeneous values of service

Mendelson and Whang [124] extended the model of §3.3, assuming m customer classes. For each $i = 1, \ldots, m$, a function $V_i(\lambda_i)$ is given, specifying the aggregate rate of utility from serving i-customers when their rate of arrival to the system is λ_i. V_i is monotone increasing, continuously differentiable, and strictly concave. V_i' can be interpreted as the value of service associated with the marginal customer, when customers are ordered in a decreasing order of service value. The service time of an i-customer is exponentially distributed with rate μ_i, and his time value is C_i per unit of time. Denote by $W_i(\underline{\lambda})$ and $W_i^q(\underline{\lambda})$ the i-customer's expected waiting and queueing time, respectively, given the vector of arrival rates $\underline{\lambda} = (\lambda_1, \ldots, \lambda_m)$. The social objective is

$$\max_{\underline{\lambda} \geq 0} \sum_{i=1}^{m} [V_i(\lambda_i) - \lambda_i C_i W_i(\underline{\lambda})].$$

For a given $\underline{\lambda}$, social welfare is maximized (under the appropriate conditions, see [52]) if the $C\mu$-rule is applied, giving higher non-preemptive priority to customers with a higher value of $C_i\mu_i$. Social optimality requires imposing prices that yield the right arrival rates under this priority rule and also induce customers to reveal their true class. Assume therefore that $W_i(\underline{\lambda})$ and $W_i^q(\underline{\lambda})$ are defined under this assumption.

The vector of optimal arrival rates, $\underline{\lambda}^*$, satisfies the first-order conditions

$$V_i'(\lambda_i^*) = C_i W_i(\underline{\lambda}^*) + \sum_{j=1}^{m} C_j \lambda_j^* \frac{\partial W_j(\underline{\lambda})}{\partial \lambda_j}\Big|_{\underline{\lambda}=\underline{\lambda}^*}, \qquad 1 \leq i \leq m. \tag{4.14}$$

If prices p_i are charged to i-customers then the equilibrium arrival rates satisfy

$$V_i'(\lambda_i) = p_i + C_i W_i(\underline{\lambda}), \quad 1 \le i \le m. \tag{4.15}$$

The following theorem is straightforward from (4.14) and (4.15).

THEOREM 4.14 *If prices p_i^*,*

$$p_i^* = \sum_{j=1}^{m} C_j \lambda_j^* \frac{\partial W_j(\underline{\lambda})}{\partial \lambda_i}\Big|_{\underline{\lambda}=\underline{\lambda}^*}, \quad 1 \le i \le m, \tag{4.16}$$

are imposed then the optimal arrival rates $\underline{\lambda}^*$ define an equilibrium.

As in the single-class case (§3.3), the value of p_i^* can be interpreted as externalities that an i-customer imposes on the others, given $\underline{\lambda}^*$.

4.3.1 Homogeneous mean service requirements

Assume that $\mu_i = \mu$ for every $i = 1, \ldots, m$, and without loss of generality, $\mu = 1$. Assume further that $C_1 > C_2 > \cdots > C_m$. Suppose that the system's organizer cannot distinguish among the classes and therefore arriving customers are asked to identify their class and are charged accordingly. The next theorem gives sufficient conditions under which customers are motivated, in equilibrium, to state their true class.

THEOREM 4.15 *Suppose that every arrival has the option to either balk or join. If he joins, he chooses one price from the set $\{p_1^*, \ldots, p_m^*\}$ defined by* (4.16). *If he pays p_i^*, he receives the priority of class i. Then, the following behavior defines an equilibrium: an i-customer joins if he values the service by more than $V_i'(\lambda_i^*)$. If he joins then he pays p_i^*.*

Proof: The $C\mu$ priority rule gives top priority to class 1, second priority to class 2, etc. From [89] p.121,

$$W_i(\underline{\lambda}^*) = W_i^q(\underline{\lambda}^*) + 1 = \frac{\sum_{j=1}^m \lambda_i^*}{S_{i-1}S_i} + 1, \quad 1 \le i \le m, \tag{4.17}$$

where $S_i = 1 - \sum_{k=1}^i \lambda_k^*$. By (4.16) and (4.17)

$$p_i^* = \sum_{j=1}^{m} \frac{\lambda_j^* C_j}{S_{j-1}S_j} + \sum_{j=1}^{m} \frac{\lambda_j^* C_j W_j^q(\underline{\lambda}^*) + \lambda_{j+1}^* C_{j+1} W_{j+1}^q(\underline{\lambda}^*)}{S_j}$$

where $\lambda_{m+1}^* = W_{n+1}^q = C_{n+1} = 0$. Let P_i^j be the full price of an i-customer who joins while claiming to be a j-customer. By (4.17)

$$P_i^j = p_j^* + C_i W_j(\underline{\lambda}^*) = p_j^* + C_i \left[\frac{\sum_{k=1}^m \lambda_k^*}{S_{j-1}S_j} + 1 \right].$$

Mendelson and Whang proved the theorem in two steps. First they proved *local optimality*, that is, for all i,

$$P_i^i \leq \min(P_i^{i-1}, P_i^{i+1}).$$

Then they proved *transitivity*: for $k > j > i$ (or $k < j < i$) if $P_i^i < P_i^j$ and $P_j^j < P_j^k$, then also $P_i^i < P_i^k$. This can be done with tedious but straightforward algebra, which we omit. □

4.3.2 Heterogeneous mean service requirements

When classes differ in their service rates, there is no analogue to Theorem 4.15. For this case, Mendelson and Whang proposed a scheme that charges customers on the basis of both the declared class and the *realized* processing time. It is assumed that upon arrival, a customer does not know his service time, but he knows to which class he belongs, and in particular his time value C_i and expected service time $\frac{1}{\mu_i}$. Let $p_i(t)$ be the amount to be paid by a customer who declares himself an i-customer and his service time turns out to be t, $0 \leq t < \infty$.

THEOREM 4.16 *There is no incentive compatible pricing scheme of the type*

$$p_i(t) = K_i f(t), \quad 1 \leq i \leq m,$$

i.e., prices that are a product of a class charge and a length-of-service charge.

THEOREM 4.17 *There is an incentive compatible pricing scheme of the type*

$$p_i(t) = A_i t + B t^2, \quad 1 \leq i \leq m.$$

Mendelson and Whang derived $A_1, \ldots, A_m$ and B. They showed that under this price mechanism each joining customer pays his conditional expected externalities (where the expectation here is over the service requirements).[10] The quadratic term in the expression can be explained by the fact that a longer service increases both the number of customers who wait for its completion and the length of time each of them waits.[11]

Mendelson and Whang actually proved a stronger property. For each (ordered) pair of classes i and k, they defined the *cheating penalty* to be the cost of announcing k when the actual class is i. Then in equilibrium, for any fixed i, this penalty function is unimodal in k with a minimum at i.

[10] The equilibrium need not be unique, see [124] p. 882. Lederer and Li [101] gave much attention to this point in their model.

[11] See Chapter 10 of Wilson's book on nonlinear pricing [176] for a broader treatment of nonlinear priority pricing.

5. Bribes and auctions

Priority pricing as used by Ghanem [55], Dolan [45], Adiri and Yechiali [2], Hassin and Haviv [72], and Mendelson and Whang [124] means that customers face a menu of priority levels from which they select one, and pay the price determined by the system manager for this priority level.

In this section we assume a different scheme in which each customer chooses the amount he wishes to pay for priority and then he is placed in the queue ahead of those who paid smaller amounts. This scheme is called *auctioning* or *bribery* according to the context. It turns out that this *decentralized* scheme also induces optimal customer behavior, i.e., a social optimal joining rate.[12]

Hassin [67] suggested auctioning as a solution to the problem of over-congestion of the equilibrium arrival rates in unobservable queues, as observed by Edelson and Hildebrand [47] (see §3.1).[13] Hassin added the following to the assumptions made by Edelson and Hildebrand:

- When joining the queue, a customer chooses a nonnegative amount to pay the server. He is not informed on the amounts paid by others.
- A customer is placed in the queue ahead of all those who paid lesser amounts. This may mean preempting the service of another customer.
- The service distribution is exponential and identical for all customers. When a customer whose service was interrupted returns to service, it is resumed from the point where it was stopped with no loss of service.
- There is no reneging, that is, a customer who joins the queue cannot leave afterwards.[14]

Let B denote an equilibrium cumulative distribution function of payments for those customers who decide to join.

LEMMA 4.18 *B is continuous and strictly increasing in an interval $[0, a]$ for some $a > 0$ and $B(a) = 1$.*

Proof: A discontinuity of B at x implies that with a positive probability there is a customer in the queue, who offers exactly x. Therefore, an

---

[12] Rashid [145] described a situation in which priorities are informally formed in a seemingly egalitarian (FCFS) system due to pressures from customers who are willing to pay in order to reduce their waiting time. Rashid concluded that the resulting corruption may increase social welfare as long as it is kept within bounds.

[13] In the general context of allocation of goods when applicants bear bid preparation costs, the decision of an individual to apply for a good may prevent another customer from receiving it. Because of this external effect the equilibrium behavior may not be socially optimal. Samuelson [151] (see [67] for extensions) showed that socially optimal participation can be induced when the allocation is done through auctioning.

[14] As observed in Remark 3.4, in the unobservable FCFS model the memoryless property of the residual waiting time makes this phenomenon a result of the model rather than an assumption.

arrival's expected welfare will be larger by a non-infinitesimal amount if he offers $x + dx$ rather than x. This contradicts the condition that in equilibrium customers maximize their expected welfare. Positivity of a follows similarly.

If $B(x)$ is constant for $b \leq x \leq d$ but increases for $x > d$, then a customer who offers b instead of d reduces his expenses without increasing the risk of getting a less favorable position. This is again a contradiction to the equilibrium requirement. The fact that B increases from 0 follows similarly. □

We will prove the optimality of the equilibrium arrival pattern in several cases. We recall that social welfare is the difference between the average rates of value obtained from service and waiting costs. The difference is a concave function of λ because the expected waiting costs are convex in λ. Therefore, to prove social optimality of the equilibrium solution it is sufficient to prove it for the marginal effect of changing the effective arrival rate.

5.1. Homogeneous customers

The equilibrium behavior of customers is characterized by a specific probability that a potential customer joins the queue and a specific distribution of payments among those who join. The expected welfare of a customer who joins equals the service value minus the payment and the expected waiting cost. In equilibrium, all customers have identical expected welfare. This value is 0 if the potential arrival rate is sufficiently large and in equilibrium some of the customers do not join.

THEOREM 4.19 *Under equilibrium, customers' joining pattern is socially optimal.*

Proof: Due to the memoryless characteristic of the exponential distribution, the residual service length of each customer in the system has the same (exponential) distribution. Also, the model assumes linear waiting costs and no loss of service due to preemption. Therefore, the order in which customers are served is irrelevant from the social point of view. This implies that the effect on social welfare that is caused by a customer who joins is independent of his payment.

In equilibrium an arriving customer is indifferent among all possible payments in $[0, a]$ for some $a > 0$. In particular, customers join as long as it is worthwhile to do so with no payment at all (i.e., with $x = 0$). By Lemma 4.18, a customer who offers no payment is placed at the end of the queue and stays there until his service is completed, imposing no extra waiting time on others.

We conclude that a customer joins if and only if he finds it worthwhile to do so when he bears all of the additional waiting costs resulting from his arrival. But this is exactly the social criterion! Thus the considerations of the individual customer and the social planner coincide, as claimed. □

We now derive further properties of the equilibrium solution. Let Λ denote the potential demand and let W_x denote the expected waiting time in equilibrium of a customer who joins the queue and pays x. We consider first the case $\lambda^* < \Lambda$. By Lemma 4.18 and the fact that in equilibrium all share the same utility, it follows that

$$x + CW_x = R, \qquad 0 \le x \le a. \tag{4.18}$$

Clearly, $W_a = \frac{1}{\mu}$. Substituting $x = a$ in (4.18) we obtain

$$a = R - \frac{C}{\mu}. \tag{4.19}$$

By (1.6), $W_0 = \frac{1}{\mu(1-\rho)^2}$. If $\Lambda > \mu - \sqrt{\frac{C\mu}{R}}$ then by Table 3.2, a customer is indifferent between joining (in particular, with zero payment) and balking, and thus $\frac{C}{\mu(1-\rho)^2} = R$. Hence, $\lambda = \mu - \sqrt{\frac{C\mu}{R}}$, which coincides with λ^* as defined in Table 3.2.

Consider now the case $\lambda^* = \Lambda$. As in (4.18), $x + CW_x$ is constant for all x, $0 \le x \le a$, but this constant may now be smaller than or equal to R. Thus, $W_x = W_0 - \frac{x}{C}$ for $0 \le x \le a$. Taking $x = a$ and recalling that $W_a = \frac{1}{\mu}$, we conclude that $a = C\left(W_0 - \frac{1}{\mu}\right) < R - \frac{C}{\mu}$, where $W_0 = \frac{1}{\mu\left(1-\frac{\Lambda}{\mu}\right)^2}$.

We now provide an alternative explanation for Theorem 4.19. The cost incurred by a customer who joins the queue is CW_0 and is independent of his payment. A customer who pays x waits only W_x, and directly contributes CW_x to the social costs. Therefore, the difference $C(W_0 - W_x)$ expresses the externalities he imposes on others. However, by (4.18) this expression equals x, so that:

- In equilibrium, the amount paid by a customer equals the externalities he imposes.

This again explains why the individual's behavior is socially optimal.[15]

[15] See the discussion in Section 4.2.

Social welfare equals

$$S = \lambda^*[R - CW(\lambda^*)], \tag{4.20}$$

where the expected waiting time, $W(\lambda^*)$, is independent of the order of service. This observation follows from the assumption of exponential service with no service loss as a result of preemption. Thus, $W(\lambda^*)$ is identical to the expected waiting time in a similar queue with an average arrival rate of λ^* and a FCFS discipline.

The server's profit consists of the payments obtained from those customers who join the queue:

$$Z = \lambda^* \int_0^a x dB(x),$$

where a is defined in (4.19). From (4.18), $x = C(W_0 - W_x)$, and hence

$$Z = \lambda^* \int_0^a C(W_0 - W_x) dB(x) = \lambda^* C[W_0 - W(\lambda^*)]. \tag{4.21}$$

Comparing (4.20) with (4.21) we find that

$$S = Z + \lambda^*(R - CW_0),$$

where the term $R - CW_0$ turns out to be the consumer surplus. Note that if not all of the potential demand is served, customers are indifferent between joining the queue and not, so that this value is 0 and then,

$$S = Z. \tag{4.22}$$

Suppose now that the assumptions of exponential service and no loss of service due to preemption are dropped, but still assume that the service demands are independent and identically distributed. The individual's decision consists of two parts: whether to join the queue, and if so, how much to pay. Hassin [67] noted that although *for a given state of the queue and a given distribution of payments* social welfare may be affected by the payment of a joining customer (certain payments may be preferred in order to avoid preemption or in order to serve a customer with a short residual service first), the *unconditional distribution* of payments is irrelevant to social welfare. The only part in the customer's decision that matters is his probability of joining.

As before, customers join as long as it is worth doing so without any payment, and this fact is independent of the service distribution. Again, the customers impose no externalities in this case and therefore, the rate by which customers join is socially optimal.

5.2. Heterogeneous customers

We now consider customers with different service values, waiting costs, or service requirements.

5.2.1 Heterogeneous service values

THEOREM 4.20 *Suppose that customers value service differently. In equilibrium, the arrival process is optimal.*

Proof: The amount of payment of a customer who joins the queue is independent of his service value. Hence, an equilibrium payment distribution for the joining customers is computed as if they came from a homogeneous population. Consequently, for some constant R_e all the joining customers have service value $R \geq R_e$ and all those who decide to balk have $R \leq R_e$.

Consider an individual who decides to balk. If he is somehow persuaded to change his decision and join, then his contribution to social welfare is $R - CW_0(\lambda)$, regardless of his payment. But this individual, while deciding not to join, already considered the possibility of joining with no payment and rejected it. Hence, $R - CW_0(\lambda) \leq 0$, and the change in his behavior does not increase social welfare. Similarly, social welfare cannot be increased by persuading customers not to join. □

5.2.2 Heterogeneous service requirements

Suppose that in addition to different service values, customers differ by their service requirements as expressed by different service rates. A customer with parameters R and μ faces the following problem:

$$\max\left\{0, \max_{x\geq 0}\left\{R - x - C\left[W_x^q(\lambda) + \frac{1}{\mu}\right]\right\}\right\},$$

where $W_x^q(\lambda)$ is the expected queueing time of a customer who pays x. By considering $R - \frac{C}{\mu}$ as the value of service completion, the payments of those who decide to join are independent of their parameters. In particular, the order in which they are served is independent of their service requirements. This is not compatible with social optimality, which requires serving customers in increasing order of their service rates.[16]

5.2.3 Heterogeneous waiting costs

We now assume that customers have different C and R values but have identical service rates. For simplicity, assume that time values C

[16]A similar difficulty exists with Mendelson and Whang's [124] priority prices, see Section 4.3.2. The solution there is to base the charges on the realized processing time.

in the population have a strictly monotone distribution function over a closed domain.

LEMMA 4.21 *Let x_1 and x_2 be payments made in equilibrium by two customers with time values C_1 and C_2, respectively. If $C_1 > C_2$ then $x_1 > x_2$.*

Proof: First we prove that $x_1 \geq x_2$. Suppose otherwise, that $C_1 > C_2$ but $x_1 < x_2$. We observe that $W(x_1) > W(x_2)$ since otherwise the C_2-customer would be better-off by reducing his payment without increasing his expected waiting time. Since the C_1-customer chose to pay x_1, it follows that

$$x_1 + C_1 W(x_1) \leq x_2 + C_1 W(x_2).$$

Since the C_2-customer chose to pay x_2, it follows that

$$x_1 + C_2 W(x_1) \geq x_2 + C_2 W(x_2).$$

Subtracting the second inequality from the first,

$$(C_1 - C_2)W(x_1) \leq (C_1 - C_2)W(x_2),$$

or $W(x_1) \leq W(x_2)$. This is a contradiction.

Suppose now that $x_1 = x_2 = x$. It follows that all customers with $C_2 \leq C \leq C_1$ also pay the amount of x. This contradicts Lemma 4.18. Thus we conclude that $x_1 > x_2$. □

THEOREM 4.22 *Suppose that customers have identical service rates, but they differ by their service values and waiting costs. Then, in equilibrium, customers' behavior is socially optimal.*

Proof: Once a customer decides to join, his payment is independent of his service value. By Lemma 4.21, individual optimization results in higher payments and priorities for the customers with the higher waiting costs, from among those who decide to join. This is clearly the socially desirable allocation of priorities among those who join. It is less obvious that the division between arrivals and non-arrivals conforms with social optimality. To demonstrate this property, assume for simplicity that there is just a finite set $C_1 < C_2 < \cdots < C_n$ of possible waiting costs. Under equilibrium, all C_i-customers who decide to join have the same expected waiting times (though not the same expected welfare since they may differ in their service value). As in Lemma 4.18, the cumulative probability distribution of payments of C_i-customers is continuous and strictly increasing on some interval $[a_{i-1}, a_i]$ with $a_{i-1} < a_i$ and $a_0 = 0$.

A C_1-customer imposes externalities only on other C_1-customers. As in Section 5.1, such a customer's payment equals the additional waiting cost he imposes on others, and a C_1-customer joins if and only if it is socially desirable for him to do so.

Suppose inductively that this property holds for C_{i-1}-customers. The external effects caused by a C_i-customer who pays a_{i-1} are the same as those caused by a C_{i-1}-customer who offers the same amount (since the service rates are identical). If he chooses to increase his payment (within the interval $[a_{i-1}, a_i]$), he will reduce his expected waiting costs but not his expected welfare (since the expected welfare is the same for all C_i-customers). Thus, the extra payment equals the expected saving in waiting costs, which in turn equals the expected additional waiting costs to other C_i-customers. We conclude that if he pays $a_{i-1} + g$ then, by the inductive assumption, a_{i-1} is the part of the externalities imposed on C_k-customers, $k < i$, and g equals the part imposed on C_i-customers. Altogether, the payment is again equal to the externalities, and a customer joins only if the sum of his waiting cost and the externalities he imposes is less then his service value. Thus, customers' equilibrium conforms with social optimization. □

5.2.4 Heterogeneous waiting costs: a special case

We now describe an explicit solution for an $M/G/1$ system with customer payments. This model was considered independently by Lui [112] and Glazer and Hassin [60]. For simplicity, we assume that the value of service is sufficiently high so that in equilibrium all of the potential demand, with rate Λ, joins.[17] Service times have a general distribution function $G(t)$ with mean $\frac{1}{\mu}$. Time value in the population is a continuous random variable with cumulative distribution $F(C)$ and density $f(C)$. The priority is assumed to be non-preemptive, but a similar analysis is possible for the preemptive case.

Suppose that a cumulative distribution of payments (or "bribes"), $B(y)$, is given. The expected cost of a C-customer who pays y, is then

$$y + C\left(W_q(y) + \frac{1}{\mu}\right) \tag{4.23}$$

where $W_q(y)$, his expected queueing time, depends on the distribution B. From Kleinrock [89] p. 12,

$$W_q(y) = \frac{W_0}{[1 - \rho + \rho B(y)]^2} \tag{4.24}$$

[17] A solution with a balking option and identical service value is presented by Lui [112].

where $\rho = \Lambda/\mu$ and $W_0 = \frac{\Lambda}{2}\int_0^\infty t^2\, dG(t)$.[18]

Recall from Lemma 4.21 that in equilibrium, the function of payments, $y(C)$ is strictly monotone increasing. The following theorem characterizes this function.

THEOREM 4.23

$$y(C) = \int_{x=0}^{C} \frac{2\rho W_0}{[1-\rho+\rho F(x)]^3} f(x)\, dx. \tag{4.25}$$

Proof: A customer wishes to minimize his expected cost (4.23). Therefore, $y(C)$ satisfies the first-order condition

$$-\frac{1}{C} = W'(y(C)).$$

This, coupled with (4.24), leads to

$$\frac{1}{C} = \frac{2W_0\rho B'(y(C))}{[1-\rho+\rho B(y(C))]^3}, \tag{4.26}$$

where $B(y)$ is the equilibrium distribution function of payment. By Lemma 4.21, $B(y(C)) = F(C)$, so that $B'(y(C)) = \frac{f(C)}{y'(C)}$. Inserting these relations in (4.26) leads to

$$y'(C) = \frac{2W_0\rho C f(C)}{[1-\rho+\rho F(C)]^3}.$$

The solution to this equation, with $y(0) = 0$ is given in (4.25). □

For the case of homogeneous customers, Glazer and Hassin [60] proved the following theorem:

THEOREM 4.24 *If all customers share the same cost parameter C, the equilibrium payment is a random variable whose support is $[0, a]$ where $a = CW_0[(1-\rho)^{-2} - 1]$ and whose cumulative distribution function is*

$$B(y) = \frac{1}{\rho}\left[\left(\frac{1}{(1-\rho)^2} - \frac{y}{CW_0}\right)^{-\frac{1}{2}} - (1-\rho)\right], \quad 0 \le y \le a. \tag{4.27}$$

[18] When priority is preemptive and service is exponential, the expected waiting time is given by

$$W(y) = \frac{1/\mu}{[1-\rho+\rho B(y)]^2}.$$

This expression differs from (4.24) by a multiplicative constant, and thus the analysis in this section applies therefore also to this case.

Proof: By (4.24)

$$y + CW_q(y) = y + \frac{CW_0}{[1-\rho+\rho B(y)]^2},$$

where $B(y)$ is the equilibrium payment distribution function. In equilibrium, $y + CW_q(y)$ is identical for all y in the support. In particular, when $y = 0$ it equals

$$\frac{CW_0}{(1-\rho)^2}.$$

The last two equations imply (4.27). The value of a is deduced from $B(a) = 1$. □

We have shown that the equilibrium solution optimizes social welfare. Another question is whether the customers gain from the institution of a payment mechanism in comparison to the FCFS discipline. We now measure customers' net welfare after subtracting their payments. Glazer and Hassin [60] checked this effect in some cases. They found that:

- If C is uniformly distributed, and in particular if all customers have the same C-value, then all customers are worse-off if a payment system is introduced.
- If there are two customer classes, the introduction of a payment system may lower the mean cost per customer.
- If C is exponentially distributed, then the introduction of the payment system does not change the mean cost per customer.

Afèche and Mendelson [3] extended the model of this section in several ways:

- Allowing a minimum price (that is, an admission price in addition to customers voluntary payment).
- Investigating the short run monopoly strategy in the heterogeneous model.
- Investigating who gains and who looses from the institution of the payment machinery in the general model.

6. Class decision

In this section we survey models about class decisions, that is, classes of customers making controlled joint decisions to maximize the welfare of the class.

Rao and Petersen [143] considered a general congestion model that has no specific queueing structure. In their model, the demand for service generates a set of classes, each associated with a demand function as well as a time value. The model assumes that there is a fixed number of priority levels, and each customer class decides how much demand to route to each priority level. The model allows the server to discriminate among customers and to impose on each of them a different cost function that linearly depends on the demand rates that the customer routes to each of the priority levels.

Given the price structure and the demand and routing decisions of the others, each customer class optimizes its own welfare. Rao and Petersen proved that this price structure is sufficiently flexible so that by selecting the right price coefficients, an equilibrium solution of customer demand and routing decisions that maximizes social welfare can be obtained. Moreover, it turns out, as in §3.1, that under these assumptions the server can extract all of the customers' surplus and thus maximizes profits by imposing socially optimal prices. To compute these prices the server must have information on the customers' demand functions and time values.

In Section 4.3 we described the model of Mendelson and Whang [124], and incentive compatible prices derived there. The authors also mentioned that these prices are not incentive compatible if each class of customers optimizes its own total welfare. They write:

> Intuitively, the class decision structure mitigates the within-class externality problem, since decisions are made by fiat for the class as a whole, but it also increases the market power of each user class. Consequently, this decision structure requires a separate analysis which is left for further research.

Van Mieghem [127] considered a single server unobservable queue with customers that send streams of *jobs* to be served. A central feature of the model is that the waiting cost is a convex function of the waiting time. It is further assumed that each customer is free to split his demand rate among priority levels, called *grades*. The rate by which customer i sends jobs to grade k is denoted by λ_k^i. The service rate for these jobs is μ^i. The service process is modeled by *classes* which capture the finest possible information that the server possesses. The server cannot distinguish among different jobs in a given class. Consequently, the scheduling rule is defined in terms of classes. Under *full information* there is a class j per each pair (i, k) and its arrival and service rates are $\Lambda_j = \lambda_k^i$ and $\mu_j = \mu^i$, respectively. Under *asymmetric information*, types are not observable and each class j consists of the aggregate demand for some service grade j, with $\Lambda_j = \sum_i \lambda_j^i$ and mean service time $\frac{1}{\mu_j} = \frac{1}{\Lambda_j} \sum_i \frac{\lambda_j^i}{\mu^i}$. The server's operating cost is convex in the aggregate demand rate.

The discipline within each grade is FCFS but the server is free to select at any moment the next customer to be served from any of the grades.

A consequence of the nonlinearity of the waiting cost function is that for any given pattern of arrival rates, a dynamic generalization of the $C\mu$-rule, denoted $GC\mu$-rule, is used to improve social welfare.

To define the $GC\mu$-rule, let C^j be the marginal delay cost of customer j, and $N_j(t)$ the number of waiting class j jobs at time t. The $GC\mu$-rule computes at time t the following index and gives priority to the class with the highest value:

$$I_j(t) = \begin{cases} C^j\left(\frac{N_j(t)}{\Lambda_j}\right)\mu_j & \text{under full information} \\ \sum_i \frac{\lambda_j^i}{\Lambda_j}\mu_j C^i\left(\frac{N_i(t)}{\Lambda_i}\right) & \text{under asymmetric information.} \end{cases}$$

Note that with full information, the index j refers to a customer whose jobs constitutes class j.

Under heavy traffic conditions it is shown by van Mieghem [126] that this rule is asymptotically optimal and reduces to the regular $C\mu$-rule if the delay cost functions are linear.

Most of the analysis of [127] relates to a model with price differentiation, that is, the queue manager can set different price menus to different customer classes. Under full information, it can extract all of the customers' surplus and, as in §3.1, profit maximization leads to a socially optimal pricing scheme. Specifically, it is shown that this can be achieved by a customer-specific two-part tariff, consisting of a fixed fee plus a variable usage fee. For asymmetric information there is a question of incentive compatibility. The results of Mendelson and Whang [124] and Lederer and Li [101] (§7.4) are extended by showing that under the dynamic $Gc\mu$-rule with arbitrary delay cost functions, there exist grade-specific prices which are socially optimal and also incentive compatible (assuming atomistic customers and homogeneous service time distributions).

Sanders [152] considered a class-decision model with an $M/M/1$ queue serving n classes. The queue manager determines the rate, λ_i, of demand by each of these classes. Class i has a utility function $U_i(\lambda_i, W)$ which is assumed to be differentiable in λ_i and W, non-decreasing in λ_i and non-increasing in W. The goal is to maximize $U = \sum_i U_i(\lambda_i, W)$ subject to $\lambda = \sum_i \lambda_i < \mu$ and $W = (\mu - \sum_i \lambda_i)^{-1}$. The problem can be solved if the utility functions are known. However, it is assumed that this information is not available. The manager can propose an allocation $\lambda_1, \ldots, \lambda_n$ and request that the classes state the values of the partial derivatives $\frac{\partial U_i}{\partial \lambda_i}$ and $\frac{\partial U_i}{\partial W}$, and update the allocation based on the reported values. This information is sufficient to apply a gradient (steepest descent) algorithm (which converges to the optimal allocation).

Classes may cheat and report partial derivatives with respect to λ_i that are larger than the true values and partial derivatives with respect to W which are smaller than the true values, in order to increase the rates allocated to them. The main result of the paper is a set of incentive compatible (possibly negative) side payments given to each class (which are functions of the reported values of all n classes). These payments induce classes to reveal their correct derivatives. This result is based on a "myopic assumption" that classes behave as if the current iteration is the final one. Thus, each class reports the value of its derivatives in order to maximize the utility associated with his side payment and the rate allocated to him in the next iteration. The incentive compatible payment suggested is a quadratic function of the reported derivatives.

Radhakrishnan and Balachandran [142] considered a firm with two divisions that share a common $M/G/1$ facility. Each division incurs a different constant waiting cost. Each division's manager determines a rate of (demand) arrival to the facility and an "effort level" which affects the profits that the division brings to the firm in a monotone increasing concave manner. The manager of the firm determines an incentive scheme of payments to the divisions, based on the realized rates of demand and profit, and aimed at maximizing the total profit of the firm. The manager of each division maximizes his own welfare consisting of the incentive payments minus costs related to the effort level and waiting costs. The actions of the divisions interact through the common expected waiting time in the queue.

The paper restricts the analysis to incentive payments that linearly depend on the realized demand and profit of the division. The main result of the paper is that a necessary condition for the incentive scheme to induce optimal behavior in equilibrium, is that the division obtains all the profits which it brings to the firm after subtracting a cost proportional to the usage rate of the facility. It is shown however, that it might be that no such scheme induces optimal demand and effort selections.

7. Related literature

- In §3.4 we observed the class dominance phenomenon which prevails in an $M/M/1$ queue: under both equilibrium and social optimization, customers from just a single class join the system (though not necessarily the same class under the two criteria). Balachandran and Radhakrishnan [18] showed that a socially optimal allocation of priorities does not necessarily give the highest priority to the class that dominates in the FCFS discipline. Another result is that there are situations where the optimal fee imposed on a higher priority class is lower than that which is imposed on a lower priority class.

- An early model of priority allocation and an incentive compatible pricing scheme is described in Marchand's work [119]. It considers an economy where individuals from a heterogeneous population consume a "composite product" and use the service provided by a facility. Their utility depends on the amount of consumption and the delay at the facility. The service and production rates are simultaneously determined so as to maximize a weighted sum of the individual benefits. It is assumed that the service requirement is deterministic and depends on customer's type. For this setting, an approximately incentive compatible pricing scheme is developed.

- Beja and Sid considered a system with a fixed arrival rate and no balking, where customers differ by their time values and service times. There are m priority classes (m is given exogenously). The manager knows the realization for each arrival and needs to assign the customer a priority class. The optimal assignment is based on $m-1$ breakpoints where the variable of interest is the cost to service time ratio (as in the $C\mu$-rule).

- Whang [174] considered a general unobservable multi-class queueing model. The utility of a class is a monotone increasing concave function of its arrival rate. Its waiting cost is a monotone increasing convex function of the total arrival rate. Whang showed that the socially optimal solution is a Pareto efficient allocation which can be achieved with various bargaining mechanisms. The mechanisms are: (1) Private bargaining for service rights; (2) Clarke tax mechanism (Section 4.2); (3) Centrally planned pricing (§3.3). Whang considered the mechanisms both under the assumption of infinitesimal customers (a "continuum economy") and in the model of class decision (a "discrete economy"). A key result of [174] is the equivalence of these mechanisms, and that the main differences among them lie with the logistics in which each of them operates and their information requirements.

- Koenigsberg [92] extended Naor's model and presented optimality conditions for profit maximization, assuming a set of heterogeneous customer classes and a preemptive non-resume priority system. However, more research is needed on this model and the resulting optimality conditions are just a first step towards understanding it.

Chapter 5

RENEGING AND JOCKEYING

This chapter discusses models in which customers react to certain conditions created *after* they join the queue. Customers may *renege* from the queue if the expected utility from remaining in the queue becomes negative. This may happen when conditions in the system deteriorate, due to a slow-down in the service rate, an increase in the expected queue length, a decrease in the value of service, or other reasons. Recall that reneging was already mentioned in §2 in a LCFS model where the last in line may renege when a new customer arrives. Another model in which customers react to changes in the state of the system deals with an observable multi-server queue where *jockeying* from one line to another (probably shorter) is allowed.

1. Reneging in observable queues

There is no individual or social incentive to renege from an observable $M/M/s$ queue if the cost of waiting per unit of time is constant, since conditions do not deteriorate over time: if it was worth joining then it is worth staying until service is completed. When the cost per unit time of waiting increases in time, reneging may be justified. A complicating issue is that it is not clear who is going to renege. A customer at the back of the queue may have a longer expected future waiting time (assuming that others in the queue will not renege) but his waiting costs per time unit are still low compared to the customers ahead of him. A distinction should be made between models that assume that each customer knows the arrival times of those ahead of him, and models in which this information is not available. In both, the customer's strategy depends on his expectation about the reneging strategies of the customers ahead of him, and an equilibrium strategy may have a complex structure.

In the following model, reneging may be exercised even when the waiting costs are linear. Assume that the service time has a decreasing hazard rate,[1] and that the length of time, t, since the current service started is known to all customers. An equilibrium strategy will be defined by thresholds $t_1 \geq t_2 \geq \cdots \geq t_n = 0$ so that a customer at position i reneges if $t \geq t_i$.

The model of Mendelson and Yechiali [125] (see also §2.10) has some similar features. They analyzed social optimality in a $GI/M/1$ system. A customer who observes i customers in the system upon his arrival balks if $i \geq n$, and enters otherwise. A customer at position n may renege if t time units have elapsed since the last arrival without any service completion. For this reason, Mendelson and Yechiali described the acceptance of the n-th customer as *conditional*. They showed that such a strategy may increase social welfare relative to the best simple control limit strategy.

The situation seems to be simpler in single-server egalitarian processor sharing (EPS) systems, where the service capacity is evenly split among the customers currently in the system. In §2.6.2 we described the way Altman and Shimkin [10] modeled the observable EPS system. Their model distinguishes the arriving customer from those already in the system by allowing him to balk after observing the state of the system. Reneging is forbidden, so the decision taken upon arrival is irrevocable.

An alternative model, treated by Assaf and Haviv [13], assumes that all customers in the system at a given time, *including a new arrival*, are indistinguishable. Therefore, it is sufficient to consider strategies that are independent of the time already spent in the system. Balking is excluded, but a customer may decide to renege at any time if this act maximizes his expected welfare, given the total number of customers present at that instant. This model is the subject of the rest of this section.

The model allows randomized strategies in which customers form lotteries continuously in order to determine whether or not to renege. The odds of these lotteries are constant as long as the number of customers in the system does not change. The resulting strategy is equivalent to

[1]For $t \geq 0$, the ***hazard rate*** of a nonnegative continuous random variable with a probability distribution F and density function f is defined as $h(t) = \frac{f(t)}{1-F(t)}$. Suppose that the waiting time is distributed according to F. Then, given that one has already waited t units of time, the probability that his waiting time terminates during the next Δ units of time is $h(t)\Delta + o(\Delta)$. F has an ***increasing hazard rate*** (IHR) if h is monotone non-decreasing in t. It has a ***decreasing hazard rate*** (DHR) if h is non-increasing in t. Without reneging, the residual waiting time in a FCFS $M/M/1$ queue is exponentially distributed, in which case the hazard rate is constant and thus it is both IHR and DHR.

one that prescribes reneging after an exponentially distributed time with a parameter chosen by the customer. The parameter may be updated when the number of customers in the system changes.

Assaf and Haviv argued that there is no (pure or mixed) equilibrium in this model.[2] First, there is no equilibrium in pure strategies since if everyone reneges at state n then an individual will do better by staying all alone and getting the full attention of the server; if n is large and no one intends to leave, then the best response for any given customer is to renege.[3] Second, there is no equilibrium in mixed strategies, since if an equilibrium prescribes mixing at state n then all customers at this state are indifferent between staying and reneging and their utility is 0. If a new arrival occurs, then the $n+1$ customers have a strictly negative utility and thus they should opt to renege immediately. But this cannot be part of an equilibrium.

Assaf and Haviv resolved the issue of the non-existence of an equilibrium by using a weaker concept of equilibrium. A strategy defines an *ϵ-equilibrium* if when all players follow it, an individual cannot increase his (expected) payoff by more than ϵ by deviating from this strategy. Assaf and Haviv showed that there exists an ϵ-equilibrium strategy, for any $\epsilon > 0$, as we describe below.

Consider an observable $M/M/1$ EPS model with time value C and service value R satisfying $R > \frac{C}{\mu}$. Denote by $(N, \theta, \eta(\epsilon))$ the strategies which are characterized by an integer threshold $N \geq 2$, a reneging rate θ, and a series of reneging rates $\{\eta_n(\epsilon)\}_{n=N+1}^{\infty}$ with $\lim_{\epsilon \to 0} \eta_n(\epsilon) = \infty$ for any $n \geq N+1$, such that:

1 No reneging occurs when the queue size is less than N.

2 Customers renege at rate θ when the queue length is N. This means that a customer reneges after an interval whose length is exponentially distributed with parameter $N\theta$, if no arrival or service completion occurs first.

3 Customers renege at rate $\eta_n(\epsilon)$ when the queue length is n, $n \geq N+1$. For small values of ϵ this implies that when the queue length reaches $N+1$, one customer (selected randomly among all present in the system) almost instantaneously reneges. Thus, states such as $N+2$ or $N+3$ are practically never reached.

[2]In fact, the argument there leads to the conclusion that no SPE exists. See Hassin and Haviv [73] for a discussion of this issue.
[3]This is an ATC situation.

The limit of $(N,\theta,\eta(\epsilon))$-strategies where the rates $\eta_n(\epsilon)$, $n \geq N+1$, approach infinity as $\epsilon \to 0$, gives the following *cooperative rule*: no one reneges when fewer than N customers are in the system; as soon as N customers are present, a reneging time is selected from an exponential distribution with parameter $N\theta$; if nobody arrives or completes service by then, then one of the customers is selected randomly to renege at that time; as soon as the number in the system reaches $N+1$, the customers form a lottery that selects one of them to leave immediately. We denote this rule as the *(N,θ)-rule*.

For $1 \leq n \leq N$ and $\theta \geq 0$, let $f_n(N,\theta)$ be the expected (future) reward for one who finds himself with $n-1$ others in the system when the (N,θ)-rule is in effect. [4] Then for $n = 1,\ldots,N-1$

$$(\mu+\lambda)f_n(N,\theta) = -C + \mu\frac{n-1}{n}f_{n-1}(N,\theta) + \lambda f_{n+1}(N,\theta) + R\mu\frac{1}{n} \tag{5.1}$$

and

$$(\mu+\lambda+N\theta)f_N(N,\theta) = -C + (\mu+N\theta)\frac{N-1}{N}f_{N-1}(N,\theta) + \frac{R\mu}{N} + \frac{\lambda N}{N+1}f_N(N,\theta). \tag{5.2}$$

Assaf and Haviv proved the following theorem:

THEOREM 5.1 *Let $N^* = \min\{N \mid N \geq 2, f_N(N,0) < 0\}$, and let $\theta^* \geq 0$ satisfy $f_{N^*}(N^*,\theta^*) = 0$. Then,*
(i) N^* and θ^* are well defined and unique.
(ii) (N^*,θ^*), coupled with some rates $\{\eta_n(\epsilon)\}_{n=N^*+1}^{\infty}$ with $\lim_{\epsilon\to 0}\eta_n(\epsilon) = \infty$ for $n \geq N^*+1$, define an ϵ-equilibrium strategy.

For any given values of N and θ, (5.1) and (5.2) are a linear system of N equations in the variables $f_1(N,\theta),\ldots,f_N(N,\theta)$. Note that $f_N(N,0)$ is monotone decreasing in N. Hence, the value of N^* can be found by solving a system as (5.1) and (5.2) with $\theta = 0$ for various values of N and using a one dimensional search for finding the smallest N with $f_N(N,0) < 0$. The next objective is to find θ^*. Using $f_{N^*}(N^*,\theta^*) = 0$ (which is a necessary condition for η – *quilibrium* for all $\eta > 0$), for the N^*-1 equations in (5.1), leads to a nonlinear system of equations with the unknowns θ^*, $f_1(N^*,\theta^*),\ldots,f_{N-1}(N^*,\theta)$. It is interesting to note that a solution can be obtained by solving an $(N^*-1)\times(N^*-1)$ system of *linear* equations, even though (5.2) contains a product of the variables θ^* and $f_{N^*-1}(N^*,\theta^*)$. See [13] for details.

[4] $f_0(N,\theta)$ need not be defined since its coefficient in (5.1) is 0.

REMARK 5.2 This section deals with observable EPS models. The unobservable EPS model has two variations. In the simpler one, customers apply strategies that do not take into account the length of time that they have been waiting. Here, the equilibrium reneging time is exponentially distributed (as one continuously forms identical lotteries between reneging and staying) and the only question left is how to compute the equilibrium parameter. In the other variation, the strategies take into account the time already spent in the system. This model seems to be more complicated.

2. Reneging in unobservable queues

Consider an unobservable queue similar to the one discussed in §3.1, but add to it the option of reneging. The queue remains unobservable after a customer joins, and in particular the customer does not know if his service has already started.

Hassin and Haviv [70] proved that the option of reneging is never exercised in equilibrium. The argument is that without reneging, the customer's virtual waiting time has a constant hazard rate. Therefore, when the option of reneging is added, the customer's virtual[5] waiting time is with IHR. Since the waiting costs per unit of time and the value of service do not vary with time, a customer who joined never reneges. In this section we describe two models in which the value of service monotonically decreases with time, and consequently reneging is exercised in equilibrium.

2.1. A single step reward function

Suppose that for some parameter T, the value of service is R if it is completed at most T time units after arrival, where $R > \frac{C}{\mu}$. Otherwise, its value is 0. Of course, no customer will stay in the system longer than T time units. Hassin and Haviv [70] proved that a customer will also never renege earlier. Again, this is due to the IHR property: while waiting, for less than T units of time, a customer's situation is constantly improving, i.e, his chances of completing waiting in the next unit of times increases with his waiting time. Hence, if he decides to join, it is better for him not to renege prior to T. A consequence of this observation is that a customer who joins, either completes service or reneges while in service, but he never reneges while waiting in the queue. The fact

[5]The ***virtual queueing time*** is defined as the time until a customer commences service assuming that he never reneges.

that customers renege affects the equilibrium joining probability, q_e. In particular, q_e is different than the one reported in Table 3.1.

Denote by (q, T) the strategy of joining with probability q and reneging T time units after joining. Hassin and Haviv proved that if everybody adopts this strategy, the expected utility of a customer who joins is

$$f(q) = \frac{R(\mu - \mu e^{-(\mu-\lambda q)T}) + \frac{\lambda p C T(\mu-\lambda q)e^{-(\mu-\lambda q)T} - C\mu(1-e^{-(\mu-\lambda q)T})}{\mu-\lambda q}}{\mu - \lambda q e^{-(\mu-\lambda q)T}}. \tag{5.3}$$

If $f(1) \geq 0$ then $q_e = 1$. Otherwise q_e is the unique root of $f(q) = 0$.

Lastly, (q_e, T) is also an ESS. When $f(1) > 0$, this is trivial. Otherwise, the best responses to (q_e, T) consist of (p, T) for any $0 \leq p \leq 1$, since the customer is indifferent between joining and balking. Yet, if all switch to p for some $p > q_e$ (which is a best response against q_e), then balking is strictly better than joining, making (q_e, T) a better response than (p, T). A similar argument holds when $p < q_e$.

2.2. Convex waiting costs

Let $c(t)$ be the waiting cost per unit of time, incurred by a customer who has already spent t time units waiting. The function $c(t)$ is assumed to be monotone increasing, reflecting reduced level of patience while waiting. Alternatively, it may represent a reduction in the value of service. Assume further that $c(0) = 0$ (implying that a customer joins upon arrival), $\lim_{t\to\infty} c(t) > \mu$ (implying reneging no later than T_2 where $c(T_2) = \mu$),[6] and that $c(t)$ is continuously differentiable.

The total cost of waiting t units of time is $C(t) = \int_0^t c(\tau)\, d\tau$. Thus, $C(t)$ is monotone increasing and convex. Assume that $R = 1$,[7] and that customers do not renege while in service. A multi-server model with these assumptions was analyzed by Haviv and Ritov [78]. In this subsection, we describe their results for the single server case.

Let $h(t)$ denote the hazard rate of the virtual queueing time. Haviv and Ritov showed that the only candidate for a *pure* equilibrium strategy is reneging at T_2. This is an equilibrium if when all customers renege at T_2, the resulting hazard function satisfies $h(t) \geq c(t)$ for every t, $t \leq T_2$. The formula of $h(t)$ is given in [78].

Haviv and Ritov proved that if

1 $\beta(t) = c(t) - \frac{c'(t)}{c(t)}$ is monotone increasing;

[6] See [78] for the analysis without these two assumptions.

[7] We could allow the service value to be time-dependent as well, but since only the ratio of these two functions matters, the assumption $R = 1$ is without loss of generality.

2 $T_2 > T_1$, where T_1 is determined by $\beta(T_1) = \mu - \lambda$;

3 $c(t)$ is concave along the interval $[T_0, T_1]$, where $T_0 < T_1$ is defined by $c(T_0) = \mu - \lambda$,

then the unique equilibrium is such that reneging time is a random variable with distribution function $G(t)$, where

$$G(t) = \begin{cases} 0 & 0 \le t < T_1 \\ 1 - \frac{\mu - \beta(t)}{\lambda} & T_1 \le t < T_2 \\ 1 & t \ge T_2. \end{cases}$$

Note that $G(t)$ defines an atom at T_2. Since $c(T_2) = \mu$, the probability mass at T_2 equals $\frac{c'(T_2)}{\lambda\mu}$.[8]

Example: Suppose that $c(t) = \sqrt{t}$. If $\lambda \le \frac{1}{2\mu^2}$ then reneging at $T_2 = \mu^2$ is the unique equilibrium. Otherwise, if $\lambda > \frac{1}{2\mu^2}$, then the unique equilibrium is

$$G(t) = \begin{cases} 0 & 0 \le t < T_1 \\ 1 - \frac{\mu - \sqrt{t} + \frac{1}{2t}}{\lambda} & t_1 \le t < \mu^2 \\ 1 & t \ge \mu^2. \end{cases}$$

2.3. Heterogeneous customers

Mandelbaum and Shimkin [117] considered an unobservable $M/M/s$ model with reneging. Once service commences, the customer is aware of this fact and he stays until his service is completed. Customers differ by their values of service R and time C, and as in the previous subsection, only the ratio $\gamma \equiv \frac{C}{R}$ matters. This ratio is continuously distributed in the population.[9] A strategy prescribes for each value of γ a (possibly random) reneging time. Since γ is random, any reneging strategy among customers leads to random reneging times across the entire population of customers.[10]

---

[8]Under other conditions, including $c(0) > 0$, there may be one more atom, at 0. This atom corresponds to the fraction of customers who balk as soon as they arrive if they find a busy server.

[9]The assumption of a continuous distribution is central. It implies that no mass of customers shares the same ratio γ.

[10]A recent paper by Shimkin and Mandelbaum [155] deals with heterogeneous customers with non-linear waiting costs.

Suppose that a γ-customer uses the strategy of reneging T_γ time units after his arrival, if he is not admitted to service by then.[11] This strategy profile defines a distribution $F(V)$ of the virtual queueing time. Consider a customer with service value R, time costs C and reneging strategy T, given the virtual waiting time distribution F. His expected net benefit is

$$U(T) = \int_0^T (R - Cv)dF(v) - CT[1 - F(T)].$$

Thus,

$$\begin{aligned} U'(T) &= (R - CT)F'(T) - C[1 - F(T)] + CTF'(T) \\ &= RF'(T) - C[1 - F(T)] \\ &= R[1 - F(T)][h(T) - \gamma], \end{aligned}$$

where $\gamma = \frac{R}{C}$ and

$$h(t) = \frac{F'(t)}{1 - F(t)}$$

is the hazard rate.[12]

In an $M/M/s$ system, the hazard rate is constant. When reneging is introduced, the virtual queueing time is with an increasing hazard rate (IHR). Therefore, U' changes its sign at most once, and when this happens the sign changes from negative to positive. Consequently, U is either monotone or unimodal with a minimum. The maximizing value is therefore in one of the extremes: $T \in \{0, \infty\}$. It follows that there is a threshold value θ such that customers with $\gamma < \theta$ never renege, and those with $\gamma > \theta$ renege immediately upon their arrival if all the servers are busy. Given that the queueing time of a customer in an $M/M/s$ system with arrival rate λ, service rate μ, and joining probability p, is positive, its distribution is exponential with parameter $s\mu - \lambda p$. Therefore, in such a system $h(t) = s\mu - \lambda p$. The threshold θ can be found by solving $\theta = s\mu - \lambda Pr[\gamma < \theta]$, where $Pr[\gamma < \theta]$ is the probability that a random customer has $\gamma < \theta$.

Mandelbaum and Shimkin also considered the same model with an additional feature: when a customer decides to arrive he is accepted only with probability q. The customer is not informed whether he has been accepted, and therefore he may keep on waiting even after being rejected. As time in the queue progresses, the prior probability q of

[11]Note that $T = 0$ means that a customer reneges immediately after arriving if the server is busy. This strategy is different from balking.

[12]In general, F has a mass at $t = 0$ and therefore $F(0)$ is defined as $F(0+)$.

having been rejected is updated. Mandelbaum and Shimkin showed the following properties:

- The posterior probability that a customer has been rejected increases with the time elapsed since his arrival.
- The queueing time has a unimodal hazard rate, first increases and then decreases. Therefore, there may be two t-values in which $h(t) = \gamma$. The intersection in the increasing phase of h corresponds to a local minimum. Hence, the best reneging strategy for a γ-customer is either $T = 0$ or where his hazard rate equals γ and is in its decreasing phase.
- There exists a unique equilibrium strategy which assigns for any γ a deterministic reneging time.

3. Jockeying

Jockeying in a multi-server queueing system is the action of moving from one queue to another. Consider two FCFS observable queues operated by identical servers, and assume that jockeying is costless. Suppose also that customers are indifferent between servers, and initially they select a server randomly. Clearly, whenever the lengths of the queues differ by 2, the last customer in the longer queue switches to the shorter one.

In contrast to this straightforward solution, when jockeying is costly the structure of an equilibrium strategy may be complex. Suppose that jockeying costs C_J. One is tempted to suggest that once the difference between the two queues is greater than $\mu C_J + 1$, the last customer in the long queue should jockey. However, jockeying under such conditions may be suboptimal. For example, if the arrival rate is very small the customer may prefer to postpone his decision until the next service completion. He may even wait for several service completions, but in such a case he takes a risk, namely that a customer in front of him in the long queue jockeys first. This model poses an interesting open questions for research.

The situation is also complex when the queues are not observable. In a common situation, a customer can observe his queue but not the other one. In such a case, a customer may try his luck and jockey even when this action is associated with a cost. Of course, he may discover that the other queue is not shorter than his original queue. It even may happen that the other queue is so long that the customer will decide to return to his original queue (bearing the jockeying cost again). The act of jockeying can be viewed in this model as that of acquiring information about the length of the other queue.

Glazer and Hassin [58] observed that under jockeying, the lengths of the two queues become positively correlated. Thus, when a customer joins a queue, he provides information about the current congestion of the other queue. This information can be used by other customers when making their decision on whether or not to jockey. This means that jockeying is associated with ***positive*** externalities between the two queues. Moreover, when other customers jockey more frequently, the differences in the lengths of the two queues decrease and the benefit an individual customer expects to gain by jockeying becomes smaller. Thus, this is an ATC situation.

Another difficulty associated with this model is that customers who have jockeyed at least once have private information about the length of the other queue.[13] This is the type of information assumed in models of retrials (see §6).

Characterizing the equilibrium strategies in models of the abovementioned type is an open research problem.

3.1. Jockeying and the value of information

Haviv and Hassin [69] considered a mixture of the observable and unobservable models of jockeying, assuming that a fraction of the customers can see the state of the system upon their arrival. The model assumes a Poisson arrival process with rate λ and two servers with separate queues and exponential service with rate μ. An arriving customer can acquire the information about which queue is shorter by paying C_I (measured in time units). A customer who does not purchase this information chooses a queue randomly. Informed customers join the shorter queue (breaking ties with equal probabilities).

Information can be acquired just once and only upon arrival. Thus, after joining, all customers are alike. Customers (informed as well as uninformed) jockey costlessly from the rear of one line to the rear of the other when the difference between them reaches a given threshold of N.[14] We assume $N \geq 3$, since information on which queue is shorter has no value when $N = 2$.

The *value of information* is the expected added time in waiting for an uninformed customer in comparison with an informed one. Let $g(p)$ be the expected value of information when the proportion of informed customers is p. Then, p defines an equilibrium if: $p = 0$ and $g(0) \leq C_I$, or $0 < p < 1$ and $g(p) = C_I$, or $p = 1$ and $g(1) \geq C_I$. When $g(p)$ is

[13] See our comment on [91] in Section 4.

[14] Without jockeying, the model is intractable, as other models in which customers join the shortest queue. See, for example, Kingman [86].

monotone decreasing, exactly one of these three cases holds and a unique equilibrium exists. However, g does not always possess this property.

One could argue that the value of information is a decreasing function of p, and conclude that this is an ATC situation: when the fraction of informed customers is large, it is less likely that a server is idle while customers are waiting in the other queue; hence, the expected waiting time is smaller. Moreover, for a fixed number of customers in the system, the expected difference in the lengths of the two queues and therefore the value of information is smaller.

The above intuitive argument is not complete. There need not be any direct relation between the expected waiting time and the reduction in expected waiting time gained when purchasing information. As a matter of fact, in numerical examples for $N = 3$, the value of information increases with p and hence FTC prevails in these cases. The explanation is as follows: for $N = 3$ the information is useful only when the difference in the lengths of the queues is exactly 1 (when it is 2, if the new arrival joins the long queue the difference will become 3 so that the customer immediately jockeys even if he does not buy the information). The probability that the difference is 1 increases with p. However, when $N \geq 4$, the numerical examples showed that the value of information decreases with p, as expected by the original intuition.

Hassin and Haviv also addressed the question of whether customers purchase the socially desired amount of information in equilibrium. If C_I is a transfer of payment and not an actual cost, then in the optimal solution everyone should be informed. If C_I is a real cost, the question of whether customers buy too much or too little information depends on whether the total gain for an individual, $g(p) - C_I$, outweighs the externalities that such an action imposes on others. Hassin and Haviv derived the externalities for $N = 3$ and found that both directions are possible in equilibrium, namely customers may buy too much or too little information than is socially desired. The following subsections are devoted to the computation of $g(p)$.

3.2. Expected waiting time

For $k \geq 0$ and $i \geq 0$, let $M_{k,i}$ be the expected (future) waiting time of a customer with k customers in front of him (in his queue), and a total of i customers behind him in his queue and from position $\max\{k-N+3, 0\}$ and up in the other queue. Note that under the model's assumption, if the length of one queue is at least $k+1$ then the length of the other queue is at least $k+2-N$. Also note that the expected waiting time of the customer under consideration is not a function of how the i customers are split between the two queues. Clearly, $M_{k,i} = \frac{k+1}{\mu}$ for $k \leq N-2$.

Without loss of generality assume that $\lambda + 2\mu = 1$. Define $M_{-1,i} = 0$, then,

$$M_{k,i} = 1 + \lambda M_{k,i+1} + \mu M_{k,i-1} + \mu M_{k-1,i+1}, \quad k \geq 0 \ , \ i \geq 1. \quad (5.4)$$

For $k \leq N-2$, $M_{k-N+1,2N-3} = \frac{k+1}{\mu}$, so that the boundary conditions are

$$M_{k,0} = 1 + \lambda M_{k,1} + \mu M_{k-1,1} + \mu M_{k-N+1,2N-3}, \quad k \geq 0,$$

$$M_{k,i} = \frac{k+1}{\mu}, \quad 0 \leq k \leq N-2, \quad i \geq 0,$$

$$\lim_{i \to \infty} M_{k,i} \leq \frac{k+1}{\mu}, \quad k \geq 0.$$

These equations and their solutions are independent of p. However, the stationary probabilities, and thus the unconditional expected waiting time and the value of information, are functions of p.

3.3. Steady-state probabilities

Hassin and Haviv applied the matrix-geometric technique for computing the stationary distribution (see Neuts [134]). Let $\pi_{i,j}$ be the stationary probability that i customers are in front of one of the servers and $j \geq i$ are in front of the other (including customers in service). For $i \geq 0$, let $L(i)$ be the set of N states $(i,i), (i,i+1), \cdots, (i,i+N-1)$. Let $\underline{\pi}_i$ be the row-vector of the stationary probabilities of the states in $L(i)$ ordered as above. For $i \geq 1$, a transition from a state in $L(i)$ can take place only to states in $L(i-1)$, $L(i)$ or $L(i+1)$. Thus, for some matrices Q_0,Q_1 and Q_2 in $\mathbb{R}^{N \times N}$ and for $i \geq 1$,

$$\underline{\pi}_i Q_0 + \underline{\pi}_{i+1} Q_1 + \underline{\pi}_{i+2} Q_2 = \underline{0}. \quad (5.5)$$

Specifically, let $\lambda_1 = \lambda\frac{1+p}{2}$ and $\lambda_2 = \lambda\frac{1-p}{2}$. Then,

$$Q_0(ij) = \begin{cases} \lambda_1 & i = 2, \ldots, N-1, \quad j = i-1 \\ \lambda & i = N, \quad j = N-1 \\ 0 & otherwise, \end{cases}$$

$$Q_1(ij) = \begin{cases} -1 & i = 1, \ldots, N, \quad j = i \\ \lambda & i = 1, \quad j = 2 \\ \mu & i = 2, \ldots, N-1, \quad j = i-1 \\ \lambda_2 & i = 2, \ldots, N-1, \quad j = i+1 \\ 2\mu & i = N, \quad j = N-1 \\ 0 & otherwise, \end{cases}$$

and

$$Q_2(ij) = \begin{cases} 2\mu & i = 1 \ , \ j = 2 \\ \mu & i = 2, \ldots, N-1 \ , \ j = i+1 \\ 0 & otherwise. \end{cases}$$

Consequently,[15] there exists a rate matrix (which is a function of p) $R \in \mathbb{R}^{N \times N}$ such that for $i \geq 0$,

$$\underline{\pi}_{i+1} = \underline{\pi}_i R.$$

Specifically, $R = \lim_{k\to\infty} X(k)$ where $X(0)$ is the zero $N \times N$ matrix, and $X(k+1) = X(k)^2 Q_2 + X(k)(I + Q_1) + Q_0$ for $k \geq 0$. By utilizing the spectral representation of the rate matrix R, an expression for the stationary probabilities can be obtained as we now describe. Let $\omega_1, \omega_2, \ldots, \omega_N$ be the eigenvalues of R. Let $E_1, E_2, \ldots, E_N$ be rank-one (projection) matrices with the properties that $E_i E_j = 0$ if $i \neq j$, $E_i E_i = E_i$, $RE_i = E_i R = w_i E_i$ for $1 \leq i, j \leq N$ and

$$R = \sum_{i=1}^{N} \omega_i E_i$$

is the spectral representation of R. Then,

$$R^k = \sum_{i=1}^{N} \omega_i^k E_i, \quad k \geq 1.$$

It was shown by Neuts [134] that when the stationary distribution exists, all eigenvalues of R are in the unit disk. Moreover, since one row of Q_0 is zero, also one row of R is zero, and hence at least one eigenvalue of R is 0. This completes the description of how to compute $\underline{\pi}_i$ for $i \geq 1$ once $\underline{\pi}_0$ is known. The straightforward details for computing $\underline{\pi}_0$ can be found in [69].

3.4. The value of information

The value of information, $g(p)$, is the difference in expected waiting times between uninformed customers and informed ones when a fraction, p, of customers purchases information:

$$\begin{aligned} g(p) &= \sum_{k=0}^{\infty} \sum_{i=1}^{N-2} \pi_{k,k+i} \left[\frac{M_{k+i,N-i-2} + M_{k,N-2+i}}{2} - M_{k,N-2+i} \right] \\ &= \sum_{k=0}^{\infty} \sum_{i=1}^{N-2} \frac{\pi_{k,k+i}}{2} \left(M_{k+i,N-i-2} - M_{k,N-2+i} \right). \end{aligned}$$

[15] See [134], pp. 80–83.

Hassin and Haviv evaluated $g(p)$ for selected values of N. They concluded that $g(p)$ is monotone increasing when $N = 3$ and monotone decreasing for larger values of N. Thus, when $N = 3$, the more customers acquire information the higher its value. Hence, if $C > g(1)$ (respectively, $C < g(0)$) nobody (respectively, everybody) acquires information and this is the unique equilibrium strategy. Moreover, it is a dominant strategy. If $g(p^*) = C$ for some $0 \leq p^* \leq 1$, then p^* prescribes an equilibrium strategy. However, in this case $p = 1$ and $p = 0$ are also equilibrium strategies.

In all the numerical examples tested by Hassin and Haviv it was shown that for $N \geq 4$ $g(p)$ is monotone decreasing, and hence if $g(0) \leq C$ then there is a unique equilibrium at $p = 0$. Moreover, not purchasing the information is a dominant strategy. The same can be said about purchasing information when $g(1) > C$. If $g(p^*) = C$ for some $p^* \in (0, 1)$ then p^* prescribes the unique equilibrium.

4. Related literature

Koenigsberg [91] considered models of two servers with reneging or jockeying. As in De Vany [41], it is assumed that the arrival rate depends only on the price (and not on the expected full price). Reneging and jockeying are done in a probabilistic way with a rate which depends on the ratio of the expected costs at the two servers. This assumption is not justified from assumptions on rational behavior of the customers. Furthermore, it is assumed that a customer knows nothing about the present state of the other system. This is problematic: a customer who just jockeyed has information about the other system. Also, the model should take into consideration that the queue lengths are not independent and therefore the length of one queue reveals information about the other one.

Chapter 6

SCHEDULES AND RETRIALS

This chapter deals with models where customers choose their time of arrival. Such models contain some non-stationary element like known times that a service facility is open, a scheduled (possibly periodic) service, or information about the state of the queue in past instants.[1]

1. Waiting time auctions

Holt and Sherman [81] considered a model for the allocation of a fixed known number, say n, of identical prizes at a specified time on a FCFS basis, to N individuals who independently choose their arrival times. To avoid trivialities assume $N > n$. Individuals are then motivated to arrive early and increase their chances of obtaining a prize. Early arrival is associated with a cost, and its benefits depend on the strategies of the other claimants. Individuals differ by their time value C and by their value R of the prize. However, only the time value of the prize $\alpha = \frac{R}{C}$ matters. It is assumed that α is continuously distributed in the population according to a distribution function G. Individuals learn whether they get a prize only at the time of allocation of prizes: "losers" also wait.

Holt and Sherman described an equilibrium in which individuals with higher time values of a prize arrive earlier. Thus, the time of arrival *before* prizes are allocated is a function $t(\alpha)$. Given $G(\alpha)$ it is possible to compute another function $F(\alpha)$, which gives the probability that an individual with time value of prize α obtains a prize. Specifically, $F(\alpha)$

[1] Related research on transportation models, originated from the work of Vickrey [170] on equilibrium travel times during the rush hours.

is the probability that the number of applicants with values greater than α, out of the other $N-1$ applicants, does not exceed $n-1$:

$$F(\alpha)=\sum_{i=0}^{n-1}\binom{N-1}{i}G(\alpha)^{N-1-i}[1-G(\alpha)]^{i}.$$

The equilibrium arrival times $t(\alpha)$ satisfy

$$t'(\alpha)=\alpha F'(\alpha) \tag{6.1}$$

for all $\alpha>0$. This relation equates the marginal values of the cost of arriving earlier and the expected value from the associated increase in the probability of getting a prize.

Assuming that there is some cost (time or monetary) associated with arriving, a threshold value α^* exists such that only those with $\alpha>\alpha^*$ choose to arrive. A person with $\alpha=\alpha^*$ is indifferent between arriving and not, and if he decides to arrive then he does so at time 0: $t(\alpha^*)=0$. With this initial condition, the solution to (6.1) is

$$t(\alpha)=\int_{\alpha^*}^{\alpha} y dF(y).$$

Holt and Sherman further verified that this is an equilibrium.

Holt and Sherman also considered variations of the model. One result is that if losers do not wait (they can observe that the length of the queue upon their arrival exceeds the number of prizes) then the equilibrium involves longer waiting times for the winners, so that the expected waiting time is unchanged.

2. $?/M/1$

Glazer and Hassin [59] considered the customer's decision of when to visit a service facility that opens daily during a prespecified time interval. They referred to the model (with a single server and exponential service) as $?/M/1$. This approach differs from most of the queueing literature which assumes that the arrival process is given exogenously. Though some papers considered the optimal setting of appointments, the abovementioned paper was the first to consider the way customers choose their arrival time in a decentralized queueing system where each customer maximizes his own welfare.[2]

The model's assumptions are:

[2]Drivers' departure time decisions in the face of travel time uncertainty have been investigated by Arnott, de Palma, and Lindsey [12] and other papers mentioned there.

1 The facility opens at time 0 and closes at time T.
2 The service discipline is FCFS.
3 All customers arriving prior to time T are served.
4 A customer may obtain a favorable position in the queue by arriving before time 0, but no service is provided until this time.
5 Customers cannot observe the queue before making their irrevocable decision of when to arrive.
6 The total number of arrivals during the day has a Poisson distribution with an expected value λ.
7 The service requirement of each customer is exponential with the same rate μ.
8 A customer has no preferences with regard to his time of arrival except that he wants to choose one that minimizes his expected waiting time.

The arrival process can be described by a distribution function $F(t)$, which gives the probability that a random customer among those who arrive during the day, arrives prior to time t. One can see (similar to §4.5) that in equilibrium the function $F(t)$ is continuous (so that no point has mass probability of arrival) and the range in which it is strictly increasing is an interval of the type $(-w, T)$ for some $w > 0$. We denote by $f(t)$ the density function corresponding to $F(t)$. Thus, the arrival process is non-homogeneous Poisson, with rate $\lambda f(t)$ at time t.

Since each customer minimizes his expected waiting time, in equilibrium each instant in which a customer may arrive has the same expected waiting time. Equivalently, since all customers have the same service distribution, customers minimize their expected queueing time. The expected queueing time of a customer who arrives at $-w$ is w, since he is guaranteed to encounter an empty queue. Therefore, in equilibrium, the expected queueing time is exactly w for an arrival at any time in the interval $(-w, T)$.

The expected queueing time for an arrival at a given instant depends on the pattern of customer arrivals prior to this instant. We will now determine necessary conditions on the function $F(t)$ that are satisfied in equilibrium.

Consider first the situation faced by a customer who arrives at time $t < 0$. The expected number of customers ahead of him in the queue is $\lambda F(t)$. Because service commences only at time zero, his expected queueing time is $w = \frac{\lambda F(t)}{\mu} - t$ (recall that $t < 0$ which explains the subtraction). Thus, $\frac{d}{dt}\left[\frac{\lambda F(t)}{\mu} - t\right] = 0$. We conclude that

- Between time $-w$ and 0, the density function is uniform., i.e.,

$$f(t) = \begin{cases} 0 & t < -w \\ \frac{\mu}{\lambda} & -w \le t < 0. \end{cases} \tag{6.2}$$

Consider next the arrival process after time 0. Let the probability that exactly k customers are in the system at time t be $P_k(t)$. Let the expected number of customers in the system at t be $N(t)$. The expected queueing time for a customer who joins at t is proportional to $N(t)$. In equilibrium, $N(t)$ is constant and $\frac{d[N(t)]}{dt} = \lambda f(t) - \mu[1 - P_0(t)] = 0$, for $0 \leq t \leq T$. Thus,

$$f(t) = \frac{[1 - P_0(t)]\mu}{\lambda}, \quad 0 \leq t \leq T. \tag{6.3}$$

Note $f(t)$ is discontinuous in 0. The rate of arrival at time t is $\lambda f(t)$, and it is independent of the realization of the arrival process prior to time t. Hence, the state probabilities $P_k(t)$ satisfy the relations

$$P_0'(t) = P_1(t)\mu - P_0(t)\lambda f(t)\ , \ 0 < t < T, \tag{6.4}$$

and for $k = 1, 2, \ldots$

$$P_k'(t) = P_{k-1}(t)\lambda f(t) + P_{k+1}(t)\mu - P_k(t)[\lambda f(t) + \mu], \quad 0 < t < T. \tag{6.5}$$

To define boundary conditions, we observe from (6.2) that the probability that a customer arrives before time 0 is $\frac{w\mu}{\lambda}$, so that the number of arrivals prior to time 0 is Poisson with mean $w\mu$, and for $k = 0, 1, 2, \ldots$

$$P_k(0) = (w\mu)^k \frac{e^{-w\mu}}{k!}\ . \tag{6.6}$$

The probability that a customer arrives after time 0 is $\int_0^T f(t)\,dt = 1 - \frac{w\mu}{\lambda}$. Therefore,

$$w = \left(1 - \int_0^T f(t)dt\right)\frac{\lambda}{\mu}. \tag{6.7}$$

The arrival pattern in equilibrium can be determined from (6.2)-(6.7). In particular, solve (6.4)-(6.5) by replacing $f(t)$ with its expression given in (6.3).

Glazer and Hassin showed that

- The arrival rate after opening time is not constant but rather decreases with t.

As in other queueing models, the individualistic behavior is not socially optimal. This problem is most clearly seen with respect to the behavior of customers prior to time 0. In equilibrium, customers arrive as early as w time units before the opening of the facility. Indeed, when λ is large relative to T, most customers join the queue before the facility opens for service. A solution to the problem of non-optimality of individualistic behavior is the institution of an appointment system, but such a solution may be too expensive in many situations. Another solution proposed in [59] is service in random order, at least among those customers in the queue at time 0. This discipline takes away the incentive to arrive prior to the opening time of the facility. Note that in such a case there is a positive probability of arrival exactly at time zero.

3. Arrivals to scheduled batch service

Glazer and Hassin [61] considered a different model in which customers decide when to arrive with the aim of minimizing their expected waiting time. In this model, service is scheduled at evenly spaced times, say $\ldots, -2, -1, 0, 1, 2, \ldots$ The interval between consecutive services is denoted as a *cycle*, and the discussion will be, without loss of generality, about the cycle $(0, 1]$. The queue discipline is FCFS and service is provided in batches of at most N customers. If more than N customers are present at the time of service then only N are served then and the rest wait for future services.[3] The arrival process is such that the arrivals in various cycles are independent and identically distributed. Specifically, let the total number of arrivals within a cycle be a Poisson random variable with mean $\lambda < N$. A customer has to decide when to arrive in his cycle.

Arrivals strictly within the cycle are socially inefficient since service does not start until the end of the cycle. Therefore, it is desired that customers arrive only at the end of the cycle. However, if there is a positive probability that the number of waiting customers exceeds N, such a solution is not an equilibrium. The reason is that by arriving infinitesimally earlier a customer may secure a better position in the queue and reduce his expected waiting time by a non-infinitesimal amount. The rest of this section characterizes the equilibrium solution.

As in the previous section, the arrival pattern is fully characterized by a probability distribution function $F(t)$ that gives the expected proportion of customers (among those who choose to come in the given cycle) who join the queue before t, $t \in (0, 1]$. In other words, $F(t)$ is the prob-

[3]Another similar situation is the "custodian's problem" described by Kosten [93]. Kosten assumed that service starts only when the server is idle and the number of waiting customers is at least a given threshold. To obtain our present model, this assumption needs to be changed to a fixed schedule.

ability that a random customer arrives at most t time units after the beginning of a cycle. Again, one easily verifies that in equilibrium, F is continuous, and the interval in which F is strictly monotone increasing has the form $(t_0, 1]$ for some $t_0 \geq 0$. The arrival process during $[t_0, 1]$ follows a non-homogeneous Poisson process with rates $\lambda(t) = \lambda \frac{dF(t)}{dt}$.

Let $W(t)$ be the expected waiting time of a customer who arrives at $t \in (0, 1]$. The equilibrium conditions imply that for some w, $W(t) = w$ for $t \in [t_0, 1]$ and $W(t) \geq w$ for $t \in (0, t_0)$. We now turn to compute w and F.

For $j = 0, 1, 2, \ldots$, let q_j be the probability that j new customers arrive in a cycle, and let r_j be the probability that j customers are in the queue just before a scheduled service. Note that the values q_j are exogenously given whereas the values of r_j are derived from the equilibrium solution. Then,

$$r_0 = q_0 \sum_{i=0}^{N} r_i \ , \tag{6.8}$$

and for $j > 0$,

$$r_j = q_j \sum_{i=0}^{N} r_i + \sum_{i=1}^{j} q_{j-i} r_{N+i}. \tag{6.9}$$

Consider a customer who chose to arrive just prior to the service at time 0. In equilibrium, he is indifferent between this choice and arriving at t_0. By definition, there are no arrivals in $(0, t_0)$. Therefore, by postponing his arrival to t_0 he does not risk his position in the queue. If the queue length prior to 0 is N or larger, then by postponing his arrival to t_0, he would save a wait of t_0. If however, the queue length prior to time 0 is at most $N-1$ then he would be served immediately if he arrives prior to 0, and if he postpones his arrival to t_0 then he would wait $1-t_0$ to be served at time 1. Therefore, $(1 - t_0) \sum_{j=0}^{N-1} r_j = t_0(1 - \sum_{j=0}^{N-1} r_j)$, or

$$t_0 = \sum_{j=0}^{N-1} r_j. \tag{6.10}$$

To complete the description of the equilibrium solution we now compute F. Denote by $P_j(t)$ the probability that exactly j customers are in the queue at time t. Then,

$$P_0(t) = \sum_{j=0}^{N} r_j \ , \quad 0 < t \leq t_0, \tag{6.11}$$

and

$$P_j(t) = r_{N+j}, \quad j = 1, 2, \ldots \quad 0 < t \le t_0. \tag{6.12}$$

The expected waiting time of a customer who arrives at time t can be expressed by

$$W(t) = (1-t) + \sum_{i=1}^{\infty} i \sum_{j=iN}^{iN+N-1} P_j(t), \quad t_0 < t < 1,$$

so that

$$\frac{dW(t)}{dt} = -1 + \sum_{i=1}^{\infty} i \sum_{j=iN}^{iN+N-1} P'_j(t) = 0, \quad t_0 < t < 1. \tag{6.13}$$

Since the arrival process is non-homogeneous Poisson,

$$P'_0(t) = -\lambda(t)P_0(t)dt, \quad t_0 < t < 1, \tag{6.14}$$

and for $j > 0$,

$$P'_j(t) = \lambda(t)[P_{j-1}(t) - P_j(t)], \quad t_0 < t < 1. \tag{6.15}$$

Substituting (6.15) in (6.13) and equating to 0 we obtain

$$\begin{aligned} 1 &= \lambda(t)\sum_{i=1}^{\infty} i\left[P_{iN-1}(t) - P_{(i+1)N-1}(t)\right] \\ &= \lambda(t)\left[\sum_{i=1}^{\infty} iP_{iN-1}(t) - \sum_{i=2}^{\infty}(i-1)P_{iN-1}(t)\right] \\ &= \lambda(t)\sum_{i=1}^{\infty} P_{iN-1}(t), \end{aligned}$$

or

$$\lambda(t) = \left[\sum_{i=1}^{\infty} P_{iN-1}(t)\right]^{-1}, \quad t_0 < t < 1. \tag{6.16}$$

Equations (6.8)-(6.16) determine the equilibrium arrival pattern. For $N = 1$, (6.16) simplifies to

$$\lambda(t) = \sum_{i=1}^{\infty}[P_{i-1}(t)]^{-1} = 1, \quad t_0 < t < 1.$$

Thus,

- When $N = 1$ the arrival rate is constant in the interval $(t_0, 1)$, and $t_0 = 1 - \lambda$.

Glazer and Hassin numerically solved the equations for other values of N and in these cases the arrival rate turned out to decrease in time within the interval $(t_0, 1)$.

4. Retrials

Another situation in which customers choose arrival times in order to maximize their welfare is concerned with *retrials* or, *repeated calls.* In these models customers who observe a busy system upon arrival, leave temporarily and return later. Between trials, the customer is said to be *in orbit.* The information that such a customer possesses is usually limited to knowledge that the servers were busy at the time of the last trial. In particular, he cannot observe the number of customers in orbit. The survey paper by Falin [49] contains an extensive discussion on this model and its variants.

In this section we consider the single server model of Kulkarni [96], Elcan [48], and Hassin and Haviv [71]. Customers arrive according to a Poisson process with rate λ, service requirements are exponential with rate μ, a customer who observes a busy server leaves temporarily and tries again later. Each retrial costs r, and the cost while orbiting is C per unit of time.[4] Balking is not allowed in this model.

Assume that the time lengths between retrials are independent and exponentially distributed with parameter η. In particular, any value of η corresponds to a strategy. Other strategies, for example when periods between retrials follow non-exponential distributions, are not allowed in this model. This assumption means that the only information that customers who are in orbit possess is that they are indeed in orbit. For example, they do not recall how many unsuccessful retrials they have had in the past or how much time has elapsed since their last retrial. In particular, they retry during the next Δt units of time with probability $\eta\Delta t + o(\Delta t)$. When we allow this probability to be time-dependent, i.e., equal to $\eta(t)\Delta t + o(\Delta t)$ for some function $\eta(t)$, where t is the time elapsed since the last retrial, corresponds to a general distribution of retrial intervals. The problem of computing an equilibrium retrial process in this more general case is still open and seems to be more complicated.

[4] Of course, only the ratio $\frac{C}{r}$ is of interest.

4.1. Steady-state probabilities

We start by deriving the state probabilities and expected time in orbit per customer (see, for example, Falin [49], and also Kulkarni [96] who gave extensions to general service distributions). The states underlying the corresponding Markov process are pairs (i, j), where i denotes whether the server is busy ($i = 1$) or not ($i = 0$), and j denotes the number of customers in orbit. Let $p_{i,j}$ be the steady-state probability of state (i, j). Then, for $i = 0, 1, \dots$

$$(\lambda + i\eta)p_{0,i} = \mu p_{1,i},$$

and

$$(i+1)\eta p_{0,i+1} = \lambda p_{1,i}.$$

Combining the two equations we obtain

$$p_{1,i+1} = \rho\left(1 + \frac{\lambda}{\eta(i+1)}\right)p_{1,i},$$

and by induction

$$p_{1,i} = \rho^i \prod_{j=1}^{i}\left(1 + \frac{\lambda}{\eta j}\right)p_{1,0}, \quad i \geq 0,$$

where an empty product is defined as 1. Consider the function

$$\begin{aligned}
G_s(x) &= \sum_{i=0}^{\infty} x^i \prod_{j=1}^{i}\left(1 + \frac{s}{j}\right) \\
&= \sum_{i=0}^{\infty} \frac{(s+1)\cdots(s+i)}{i!}x^i \\
&= \sum_{i=0}^{\infty} \binom{s+i}{i} x^i \\
&= (1-x)^{-1-s}.
\end{aligned}$$

Then, using $\sum_{i=0}^{\infty} p_{1,i} = \rho$ (which is the fraction of time in which the server is busy) and substituting $x = \rho$ and $s = \frac{\lambda}{\eta}$ we conclude that

$$p_{1,0} = \rho\left[\sum_{i=0}^{\infty}\rho^i \prod_{j=1}^{i}\left(1 + \frac{\lambda}{\eta j}\right)\right]^{-1} = \rho\left[G_{\frac{\lambda}{\eta}}(\rho)\right]^{-1} = \rho(1-\rho)^{1+\frac{\lambda}{\eta}}. \quad (6.17)$$

It follows that for $i \geq 0$,

$$p_{0,i} = \rho^{i+1}(1-\rho)^{1+\frac{\lambda}{\eta}}\frac{\mu}{\lambda + i\eta}\prod_{j=1}^{i}\left(1 + \frac{\lambda}{\eta j}\right).$$

We next compute the expected number of customers in orbit. First,

$$\sum_{i=1}^{\infty} ip_{1,i} = p_{1,0} \sum_{i=1}^{\infty} i\rho^i \prod_{j=1}^{i} \left(1 + \frac{\lambda}{\eta j}\right) = p_{1,0} \left(1 + \frac{\lambda}{\eta}\right) \rho \left(1 - \rho\right)^{-\left(\frac{\lambda}{\eta} + 2\right)},$$

where the second equality follows from the observation that $\frac{\partial}{\partial x} G_{\frac{\lambda}{\eta}}(x)|_{x=\rho} = (1 + \frac{\lambda}{\eta})(1-\rho)^{-2-\frac{\lambda}{\eta}}$. Combining with (6.17) we obtain

$$\sum_{i=1}^{\infty} ip_{1,i} = \frac{\rho^2}{1-\rho} \left(1 + \frac{\lambda}{\eta}\right). \tag{6.18}$$

Similarly,

$$\sum_{i=1}^{\infty} ip_{0,i} = p_{1,0} \frac{\lambda}{\eta} \left(1 - \rho\right)^{-\left(1 + \frac{\lambda}{\eta}\right)},$$

which, with (6.17), gives

$$\sum_{i=1} ip_{0,i} = \frac{\lambda}{\eta} \rho. \tag{6.19}$$

By (6.18) and (6.19), the expected number of customers in orbit is

$$L_q = \sum_{i=1}^{\infty} i(p_{0,i} + p_{1,i}) = \frac{\rho^2}{1-\rho} + \frac{\lambda \rho}{(1-\rho)\eta}. \tag{6.20}$$

L_q equals the queue length in the corresponding $M/M/1$ queue, plus a term which is the expected *excess* of customers in the system in this model.[5] As expected, this excess is monotonically decreasing with η. Moreover, it approaches 0 when $\eta \to \infty$. Since the arrival rate to the orbit queue is $\lambda\rho$, we obtain by Little's formula that the expected time in orbit for a customer who observes a busy server upon his initial arrival is

$$\frac{1}{1-\rho} \left(\frac{1}{\mu} + \frac{1}{\eta}\right).$$

Note that his (conditional) expected waiting time is decomposed into two sources. The first is as in a standard $M/M/1$ queue, and the second term expresses therefore the expected time added because of the non-work-conserving nature of the model. This decomposition is typical in models that involve *server vacations*.

[5]The model is not work-conserving since it allows the server to stay idle while customers are waiting to be served.

4.2. Social optimality

Kulkarni [95] computed the socially optimal rate of retrials. The problem is to select a value for η which minimizes $CL_q + r\eta L_q$. An increase in η is associated with more frequent retrials. This means retrying when the probability that the server is still busy increases, and therefore an increase in η leads to an increase in the expected number of retrials. On the other hand, a reduction of η is associated with an increase in the expected time a customer spends in orbit. The optimal rate balances these two effects.

Substituting L_q from (6.20) and differentiating with respect to η, we obtain from the first-order condition for the optimal rate, that

$$\eta^* = \sqrt{\frac{C\mu}{r}}.$$

We observe that:

- The optimal retrial rate is independent of the arrival rate.

An increase in λ increases the utilization of the server (as expressed by λ/μ). Thus, for any value of η, both the expected number of retrials and the expected time in orbit increase. The surprising result is that these changes preserve the optimal balance between the two opposing effects and thus the value of η^* is not affected.

4.3. Equilibrium

We now derive the equilibrium retrial rate for the case of exponential service as given by Elcan [48]. The derivation for an arbitrary service distribution appears in Hassin and Haviv [71] and is based on results from Kulkarni [95].

We need to derive the expected orbiting time for a tagged customer who uses a retrial rate of γ when all others use η. We denote this value by $g(\gamma, \eta)$. Note that we do not assume in this definition that the customer observes a busy server upon arrival (and hence, the time in orbit may be 0 with a positive probability).

The expected total cost for such an individual is

$$f(\gamma, \eta) = C\left(g(\gamma, \eta) + \frac{1}{\mu}\right) + \gamma r g(\gamma, \eta).$$

In equilibrium, if all use η then this is also a best response. Hence, we look for an η such that

$$\left.\frac{\partial f(\gamma, \eta)}{\partial \gamma}\right|_{\gamma=\eta} = 0. \qquad (6.21)$$

We will show that such a rate exists and that it is unique.

Let ϕ_i denote the expected time in orbit of the tagged customer given that he is in orbit, the server is busy, and there are i other customers in orbit. Then, $g(\gamma,\eta) = \sum_{i=0}^{\infty} p_{1,i}\phi_i$.[6] The following lemma appears in [48]. We give an alternative proof.

LEMMA 6.1 *For $i \geq 0$,*

$$\phi_i = ai + b$$

with

$$a = \frac{\eta}{\mu[(1-\rho)\eta + \gamma]}, \tag{6.22}$$

and

$$b = \frac{1}{\mu} + \frac{1+\rho}{\gamma} + \frac{\rho\eta(\lambda+\gamma)}{\mu\gamma[(1-\rho)\eta+\gamma]}. \tag{6.23}$$

Proof: We first argue for the affinity of ϕ_i as a function of i. Compare the tagged customer when he is with i or $i+1$ others in orbit. The extra customer in the latter case inflicts an expected added time in orbit on the tagged customer. We argue that this added value is independent of i. While in orbit, the tagged customer's retrials form a Poisson process with rate γ. Likewise, the retrials of the extra customer form a Poisson process with rate η. Therefore, the extra customer enters service first, and hence increases the delay of the tagged customer, with probability $\frac{\eta}{\eta+\gamma}$ (which is independent of i). This expected added delay equals the expected service time of the extra customer, $\frac{1}{\mu}$, plus service times of those who arrive during his service time and who overtake him, plus the service times of those who arrive and overtake the tagged customer while the latter are served, and so on. Again, all these values are independent of i. This completes the proof for the affinity of ϕ_i.

Note that a is the expected added delay of the tagged customer due to the above-mentioned extra customer. This added delay is 0 with probability $\frac{\gamma}{\gamma+\eta}$ and positive with probability $\frac{\eta}{\gamma+\eta}$. Conditioning on the latter case, the expected delay equals the expected service time $\frac{1}{\mu}$ plus the expected number of customers to arrive during this service time, $\frac{\lambda}{\mu}$, multiplied by the expected delay each of them adds to the tagged customer, which is again a. In summary,

$$a = \frac{\eta}{\gamma+\eta}\left(\frac{1}{\mu} + \frac{\lambda}{\mu}a\right),$$

[6] $p_{1,i}$ is a function of η and ϕ_i is a function of γ and η. In order to simplify the presentation, we omit reference to these parameters.

which is solved uniquely by the value for a stated in (6.22).

Lastly, b is the expected time in orbit for the tagged customer who is orbiting alone given that the server is busy. Thus,

$$b = \frac{1}{\lambda+\mu} + \frac{\lambda}{\lambda+\mu}(a+b) + \frac{\mu}{\lambda+\mu}W \tag{6.24}$$

where W is the expected waiting time for the tagged customer who is orbiting alone while the server is idle. Also,

$$W = \frac{1}{\lambda+\gamma} + \frac{\lambda}{\lambda+\gamma}b. \tag{6.25}$$

Solving (6.24) and (6.25) for b, completes the proof. □

THEOREM 6.2 *A unique equilibrium retrial rate η_e exists, with*

$$\eta_e = \frac{C\rho + \sqrt{C^2\rho^2 + 8\mu Cr(1-\rho)(2-\rho)}}{4r(1-\rho)}.$$

Proof: Lemma 6.1 and (6.18) imply that

$$\begin{aligned} g(\gamma,\eta) &= \sum_{i=0}^{\infty}(ai+b)p_{1,i} \\ &= a\sum_{i=0}^{\infty} ip_{1,i} + b\rho \\ &= a\frac{\rho^2}{1-\rho}\left(1+\frac{\lambda}{\eta}\right) + b\rho. \end{aligned}$$

Substituting a and b from (6.22) and (6.23), we obtain

$$g(\gamma,\eta) = \frac{\rho}{1-\rho}\left[\frac{1}{\mu} + \frac{1}{\gamma} + \frac{\rho}{\mu}\frac{\eta-\gamma}{(1-\rho)\eta+\gamma}\right]. \tag{6.26}$$

Therefore,

$$\left.\frac{\partial f(\gamma,\eta)}{\partial\gamma}\right|_{\gamma=\eta} = \frac{\rho}{1-\rho}\left[r\left(\frac{1}{\eta}+\frac{1}{\mu}\right) + (C+r\eta)\left(-\frac{1}{\eta^2} + \frac{\rho}{\mu}\frac{1}{\eta(\rho-2)}\right)\right].$$

With (6.21), this gives a quadratic function

$$2r(1-\rho)\eta^2 - C\rho\eta - C\mu(2-\rho) = 0,$$

which has a unique positive root, η_e, as claimed. □

We note that

- η_e increases monotonically with λ.
- When $\rho \uparrow 1$, $\eta_e \uparrow \infty$.
- When $\rho \downarrow 0$, $\eta_e \downarrow \eta^*$.
- The equilibrium retrial rate is higher than the optimal retrial rate. This property follows since $2 - \rho > 2(1 - \rho)$ so that

$$\eta_e > \frac{\sqrt{C\mu r(1-\rho)^2}}{r(1-\rho)} = \eta^*. \tag{6.27}$$

  In particular, in equilibrium customers stay in orbit less time than they do under social optimization. Yet, of course, their total costs are higher.

- When a customer selects his retrial rate he ignores the way this act affects other customers. Thus, the optimal and equilibrium retrial rates differ. When a customer in orbit retries he may prevent another customer, new or in orbit, from succeeding. This means that an arrival, and in particular a retrial, is associated with negative externalities. We observed that $\eta_e > \eta^*$. This reflects the fact that the length of the queue t time units after the server was observed to be busy, stochastically decreases with t. Therefore, the negative externalities associated with a retrial decrease in the time since the previous trial. Hence, a customer who ignores these externalities retries sooner than is socially desired.

- The expected cost of a customer increases as others retry at a higher rate. In other words, $f(\gamma, \eta)$ increases monotonically with η for any fixed γ. To verify this property note that $f(\gamma,\eta) - \frac{C}{\mu} = (C + \gamma r) g(\gamma, \eta)$ so that $\frac{\partial f(\gamma,\eta)}{\partial \eta}$ is proportional to $\frac{\partial g(\gamma,\eta)}{\partial \eta}$. Using (6.26), these derivatives have the same sign as

$$(1-\rho)\eta + \gamma - (\eta - \gamma)(1-\rho) = \gamma(2-\rho)$$

  and therefore it is positive. An intuitive explanation for this property can be found on lines similar to the arguments for the previous observation. The higher the retrial rates of others, the higher the probability that the server is busy again when the tagged customer retries.

- Equation 34 of Elcan [48] gives the derivative of an individual's best response γ as a function of the rate η adopted by the others. When $\eta = \eta_e$ this derivative is negative, which shows that in the neighborhood of η_e the situation is ATC.

Hassin and Haviv [71] suggested two ways to resolve the difference between the optimal and equilibrium retrial rates. The first is by partial compensation (a *rebate*) to customers for their actual waiting costs. The other is by taxing unsuccessful retrials.

THEOREM 6.3 *A positive rebate, $P < C$, per unit of waiting time exists, that induces an equilibrium with the optimal retrial rate η^*. It is determined by*

$$\frac{(C-P)\rho + \sqrt{(C-P)^2\rho^2 + 8(C-P)\mu r(1-\rho)(2-\rho)}}{4r(1-\rho)} = \sqrt{\frac{\mu C}{r}}.$$

The same effect can be achieved by imposing a toll $T = r\frac{P}{C-P}$ per retrial so that the cost per retrial is now $r + T$.

Proof: By (6.27), $\eta_e > \eta^*$. Also, note that η_e is continuously monotone increasing in C. Therefore, it suffices to show that for sufficiently small values of C, the resulting equilibrium retrial rate is smaller than the original η^*. Indeed, when $C - P$ approaches 0, the value of η_e also approaches 0 and therefore, for some intermediate value of P, equality holds. In particular, the stated equation has a unique solution $P \in (0, C)$.

To compute T observe that the optimal and the equilibrium rates depend on r and C only through their ratio. We obtain the desired toll value by solving $\frac{r+T}{C} = \frac{r}{C-P}$. □

5. Related literature

- Daniel [39] considered a model in which a given set of identical customers would most like to be served at a specific time. In addition to the usual linear waiting costs, customers also incur linear costs if service starts earlier or later than this desired point in time. The number of customers is Poisson and the length of service is deterministic. Each customer announces his scheduled arrival time but the actual arrival time may turn out to be different since the actual time of arrival is subject to independent random shocks. The paper gives conditions for equilibrium scheduling as well as conditions for socially optimal scheduling. The social planner can achieve a decentralized implementation of the optimal arrival schedule by imposing a congestion fee equal to the externality cost. If customers' declared scheduled times of arrival cannot be trusted, the fees depend on the actual times of arrival. Daniel presented an algorithm for computing an equilibrium solution based on the observation that the expected cost must be identical in each point of time used to schedule an arrival.

The paper was motivated by the regulation of aircraft schedules and also deals with the case in which many of the arrivals belong to the same airline, which then internalizes some of the external costs.

- Rapoport, Stein, Parco, and Seale [144] considered a model in which customers choose their arrival times to a service facility. The facility is open during a given time interval. It is not possible to queue before opening and a customer whose service is not completed before closing time is not served. The queue is unobservable so that an arriving customer always joins and then stays until either his service is completed or the facility closes, whichever comes first. The paper contains numerical solutions for the arrival pattern in equilibrium for selected sets of parameters, and reports on an experimental study with 20 players.

Chapter 7

COMPETITION AMONG SERVERS

This chapter deals with markets in which servers compete over the customers, usually by posting prices. Most of the models consider a game with two stages; servers act as leaders by announcing prices, and customers follow by selecting servers accordingly. Thus, the model computes customers' equilibrium for any given set of prices, so that each customer optimizes his own welfare by choosing a server. Then, an equilibrium among the servers is computed, where each server sets a price that maximizes its profits, given the prices of the others. At this stage, the servers assume that for each set of prices, the arrival rates are determined by the corresponding customers' equilibrium.

Competition in models with congestion has a unique feature which Reitman [146] points out: firms that sell the same product still have positive profits. The reason is that by cutting its price below the market price, a server is still unable to attract all of the customers from its competitors: as its demand increases, waiting times increase and thus the quality of the service it offers decreases. The number of customers who switch to a server that cuts its price is limited by the accompanying deterioration in its quality. In equilibrium, if the price is reduced then the gain from new customers equals the revenues lost from existing customers.

1. Unobservable queues with heterogeneous time values

1.1. Continuous distribution of time values

The earliest works on competition between servers in a queueing model are Luski [113] and Levhari and Luski [104]. The two papers deal with the same model and introduce complementary results.

Consider two identical exponential servers with separate unobservable queues. The joint arrival process is Poisson with intensity λ. Customers differ by their time value C which has a continuous distribution function $F(C)$.[1] The value of service, R, is identical for all customers.

Suppose that server i charges an admission fee of p_i, $i = 1, 2$. Without loss of generality, assume that $p_1 \geq p_2$. Also assume that each server serves a positive fraction of the population. This situation is possible in equilibrium only if the corresponding expected waiting times, W_1 and W_2, satisfy $W_1 \leq W_2$. In this case, customers with lower C values join server 2, those with higher C values join server 1, and those with highest C values may balk. In particular, a customer with time value C joins server 2 if $CW_2 + p_2 \leq CW_1 + p_1$ (equivalently, $C < \frac{p_1 - p_2}{W_2 - W_1}$), balks if $C > \frac{R - p_1}{W_1}$, and joins server 1 if $\frac{p_1 - p_2}{W_2 - W_1} < C < \frac{R - p_1}{W_1}$.

This means that the arrival rates to the servers satisfy

$$\lambda_2 = \lambda F\left(\frac{p_1 - p_2}{W_2 - W_1}\right),$$

and

$$\lambda_1 = \lambda\left[F\left(\frac{R - p_1}{W_1}\right) - F\left(\frac{p_1 - p_2}{W_2 - W_1}\right)\right] = \lambda F\left(\frac{R - p_1}{W_1}\right) - \lambda_2.$$

Substituting $W_i = \frac{1}{\mu - \lambda_i}$, we obtain for any given pair of prices p_1 and p_2 a system of two nonlinear equations with variables λ_1 and λ_2,

Each server wishes to maximize profits, $p_i \lambda_i$ $i = 1, 2$. Levhari and Luski [104] solved several instances. In each case they computed the *reaction curve* which describes the profit-maximizing price of a server given the price set by the other server. From the reaction curve one can deduce the equilibrium solutions. Their main results are:

---

[1]This assumption implies that there is no C value common to a positive fraction of the population of customers or, put differently, no two customers share the same cost parameter C. The other extreme, in which the value of C is common to all customers, was treated by Chen and Wan [35]. The case of only two possible values for C was treated by Armony and Haviv [11]. See Section 1.2.

- In some cases an equilibrium comes with equal prices, in others the prices differ. However, the social welfare maximizing prices are always different. The latter result was proved analytically.[2]
- Whenever the prices differ in equilibrium, the profits also differ.
- Luski and Levhari also computed the profit-maximizing prices set by a monopoly that owns both servers, and the social welfare maximizing prices. They report that in numerous examples the prices set at the two servers differ, and are higher than those obtained in the case of competing servers.
- The social welfare maximizing prices are lower than the equilibrium prices.

Reitman [146] generalized the results of Luski and Levhari, by considering a multi-server model with identical servers and heterogeneous customers. Server i, $i = 1, \ldots, n$, chooses an admission fee p_i and a service capacity μ_i, and consequently faces an arrival process with rate λ_i. The waiting time at the server is some strictly increasing function of $\frac{\lambda_i}{\mu_i}$. Customers differ by their time value.

Reitman proved that when $n \geq 3$ only asymmetric equilibria are possible. In equilibrium, customers with lower time values choose lower priced and higher congested servers. Therefore, there exist thresholds $C_0 = 0 < C_1 < \cdots < C_{n-1} < C_n = \infty$ such that customers with time values $C_{i-1} \leq C < C_i$ choose server i, $i = 1, \ldots, n$.[3] Since the servers' profits are positive, if no fixed costs are entailed, new firms join the market and the number of servers become infinite. In this case, all firms charge different prices in equilibrium.

1.2. Two time values

Armony and Haviv [11] considered a model that differs from the model of Levhari and Luski [104] by one key distinction: a customer's time value is either C_1 or C_2, where $C_1 < C_2$.

Consider two identical exponential servers with separate unobservable queues. The joint arrival process is Poisson with intensity λ. Customers are homogeneous with respect to their value of service. Their cost parameters are C_1 or C_2 with probabilities q_1 and $q_2 = 1 - q_1$, respectively. Joining queue i, comes with a charge of p_i, $i = 1, 2$.

---

[2] Loch [111] proved that when the time values of all customers are identical, the equilibrium prices are symmetric.

[3] Compare with §4.4.1.

The game has two levels. In the first level, there is a game among customers who observe the prices set by the servers. Each customer chooses between joining one of the servers or balking. Armony and Haviv observed that unless the prices are equal, there exists a unique equilibrium. An equilibrium comes with a joining/balking *pattern*, namely the set of actions that have positive probabilities, for each of the two customer types.

Armony and Haviv suggested an algorithm for computing an equilibrium for a given set of prices. It initiates with a particular equilibrium for the case in which the two prices are equal to the lower of the two given prices, and increases one of the prices until the higher of the given prices is reached. At certain values the equilibrium pattern changes. Armony and Haviv showed how to compute the resulting pattern progression. They observed a surprising possibility: the total arrival rate to the system may increase when the prices increase.

At the second level, there is a game between the servers. For any given price of one of the servers, Armony and Haviv used the abovementioned algorithm to compute the best response price for the other. This leads to a *reaction curve* which is the function of a server's best response price. This curve is used to determine equilibrium prices. By running numerical examples, Armony and Haviv concluded that the following four cases are possible: a unique symmetric equilibrium, multiple asymmetric equilibria, a continuum of non-symmetric equilibrium, or no (pure) equilibrium exists.

2. Unobservable queues with heterogeneous values of service

Loch [111] considered price competition between two identical $M/G/1$ servers of unobservable queues who serve customers with heterogeneous service values. We describe Loch's result for the case of two identical servers. The assumptions concerning the customers are similar to those postulated by Mendelson and Whang [124] (§4.4.3): there exists an increasing concave utility function $V(\lambda)$ which represents the total customers' expected rate of utility from service when the arrival rate is λ. Denote $V'(\lambda)$ by $P(\lambda)$. To avoid trivialities, i.e., an analysis which ends up with no arrivals, assume that $P(0)$ is sufficiently large. Under the equilibrium arrival rate λ, the marginal utility $P(\lambda)$ equals the expected full price at any active server.

2.1. Single class of customers

Loch considered several models with two types of equilibria and two types of optimization problems.

- **Bertrand equilibrium (price competition)** Suppose that the servers first select prices p_1 and p_2, and then the arrival rates λ_1 and λ_2 are determined. Server 1 (and similarly for server 2) takes p_2 as given and sets p_1 to maximize $\lambda_1 p_1$ under the customers' equilibrium conditions

$$P(\lambda_1 + \lambda_2) = p_1 + CW(\lambda_1) = p_2 + CW(\lambda_2), \tag{7.1}$$

  where W denotes the expected waiting time. Loch showed that in equilibrium $p_1 = p_2$ and hence $\lambda_1 = \lambda_2 = \lambda$, where the common arrival rate λ satisfies

$$P(2\lambda) + \lambda P'(2\lambda)\frac{CW'(\lambda)}{CW'(\lambda) - P'(2\lambda)} = C[W(\lambda) + \lambda W'(\lambda)].$$

- **Cournot equilibrium (rate competition)** Suppose now that the servers choose arrival rates λ_1 and λ_2, and prices are determined by the equilibrium conditions (7.1). Loch showed that in equilibrium $\lambda_1 = \lambda_2 = \lambda$, where the common rate λ satisfies

$$P(2\lambda) + \lambda P'(2\lambda) = C[W(\lambda) + \lambda W'(\lambda)].$$

- **Monopoly** Suppose that the two servers are owned by a monopoly. The monopoly selects prices p_1 and p_2 to maximize $\lambda_1 p_1 + \lambda_2 p_2$, subject to the equilibrium condition (7.1). This is equivalent to

$$\max_{\lambda_1, \lambda_2 \geq 0} \sum_{i=1}^{2} \lambda_i [P(\lambda_1 + \lambda_2) - CW(\lambda_i)]$$

  subject to

$$P(\lambda_1 + \lambda_2) - CW(\lambda_i) \geq 0, \quad i = 1, 2. \tag{7.2}$$

  Since W is a convex function, the objective function is concave and only first-order conditions need to be considered. Assuming an interior solution, i.e., positive arrival rates and equality in (7.2), for $i = 1, 2$,

$$P(\lambda_1 + \lambda_2) + (\lambda_1 + \lambda_2)P'(\lambda_1 + \lambda_2) - C[W(\lambda_i) + \lambda_i W'(\lambda_i)] = 0.$$

  Since $W(\lambda_i) + \lambda_i W'(\lambda_i)$ is strictly monotone in λ_i, it follows that $\lambda_1 = \lambda_2$.

- **Maximization of social welfare** The objective is to maximize $V(\lambda_1 + \lambda_2) - \sum_{i=1}^{2} C\lambda_i W(\lambda_i)$. This is a concave function as in the previous case. The first-order conditions are

$$P(\lambda_1 + \lambda_2) = C[W(\lambda_i) + \lambda_i W'(\lambda_i)], \quad i = 1, 2.$$

Again, in the optimal solution, $\lambda_1 = \lambda_2 = \lambda$. These arrival rates can be induced by imposing prices $p_1 = p_2 = C\lambda W'(\lambda)$.

Loch proved the following theorem:

THEOREM 7.1 *Let λ_B be the total arrival rate under Bertrand equilibrium, λ_C under Cournot equilibrium, λ_M under monopoly optimization, and λ_S under social optimization. Then,*

$$\lambda_S > \lambda_B > \lambda_C > \lambda_M.$$

2.2. Multiple classes of customers

Loch also investigated the Cournot equilibrium in a model with two identical servers and n classes of customers. Class k is characterized by a time value C_k, a service rate μ_k, a function V_k of aggregate benefit from service, and a marginal utility function $P_k = V_k'$ with a sufficiently large values of $P_k(0)$, $1 \le k \le n$, to guarantee that $\lambda_k > 0$ in equilibrium. Customers receive priority according to the $C\mu$-rule.

Suppose that for $i = 1, 2$ and $k = 1, \ldots, n$, the arrival rate of k-customers to server i is $\lambda_{ik} > 0$, and let p_{ik} be the corresponding admission fee. Given $(\lambda_{j1}, \ldots, \lambda_{jn})$, for $j \neq i$, server i selects nonnegative rates $(\lambda_{i1}, \ldots, \lambda_{in})$ that maximize $\sum_{k=1}^{n} \lambda_{ik} p_{ik}$ subject to the equilibrium conditions

$$P_k(\lambda_{1k} + \lambda_{2k}) = p_{ik} + C_k W_k(\lambda_{i1}, \ldots, \lambda_{in}), \quad i = 1, 2, \quad k = 1, \ldots, n,$$

where $W_k(\lambda_1, \ldots, \lambda_n)$ is the expected waiting time of a k-customer who joins a queue when the arrival rates to this server are $(\lambda_1, \ldots, \lambda_n)$.

Loch proved that there exists an equilibrium such that $\lambda_{1k} = \lambda_{2k}$ for $k = 1, \ldots, n$. Denoting this common rate by λ_k, the first-order optimality conditions are: for $k = 1, \ldots, n$,

$$P_k(2\lambda_k) + \lambda_k P_k'(2\lambda_k) = C_k W_k(\lambda_1, \ldots, \lambda_n) + \sum_{l=1}^{n} C_l \lambda_l \frac{\partial W_l(\lambda_1, \ldots, \lambda_n)}{\partial \lambda_k}. \quad (7.3)$$

Under social optimization, the objective is to set $\lambda_1, \ldots, \lambda_n$ to maximize $\sum_{k=1}^{n} [V(2\lambda_k) - 2C_k \lambda_k W_k(\lambda_1, \ldots, \lambda_n)]$. This objective function is based on the fact that under social optimization each class evenly splits

its arrival rates among the two servers. The first-order optimality conditions are: for $k = 1, \ldots, n$,

$$P_k(2\lambda_k) - C_k W_k(\lambda_1, \ldots, \lambda_n) - \sum_{l=1}^{n} C_l \lambda_l \frac{\partial W_l(\lambda_1, \ldots, \lambda_n)}{\partial \lambda_k} = 0.$$

Combining with (7.3), the prices which induce social optimality are

$$p_k = \sum_{l=1}^{n} C_l \lambda_l \frac{\partial W_l(\lambda_1, \ldots, \lambda_n)}{\partial \lambda_k} - \lambda_k P_k'(2\lambda_k), \quad 1 \le k \le n, \tag{7.4}$$

where $\lambda_1, \ldots, \lambda_n$ are the socially optimal arrival rates.

As in §4.4.3, Loch showed that also in this model, the pricing scheme (7.4) is not incentive compatible and customers may have an incentive to lie about their type. However, it is possible to impose incentive-compatible prices which depend on the *actual* service requirement and the customer *claimed* type.

3. Observable queues

When servers compete over identical customers and the queues are observable, there is no price equilibrium. Suppose that there are two servers, server 1 and server 2, who set admission fees p_1 and p_2, respectively, such that $p_2 \ge p_1 \ge 0$. Suppose, for simplicity, that balking and jockeying are not possible, and therefore only the difference $\delta = p_2 - p_1$ matters. Consider first the case where $\delta = m\frac{C}{\mu}$ for some integer m. With positive probability, a customer observes upon arrival that the queue at server 1 is longer than the queue at server 2 by exactly m customers. In this case, the customer is indifferent between the servers and selects one of them by some tie-breaking rule. This situation is not sustainable in equilibrium since at least one of the servers can increase profits by an infinitesimal price reduction.[4] Consider now the other case, where $\delta \ne m\frac{C}{\mu}$ for every nonnegative integer m. This situation is also not sustainable since each of the servers can now increase profits by a small price increase that does not alter the selection strategy of the customers. Therefore, no equilibrium exists. A similar argument shows that there is no equilibrium also when servers can set state-dependent fees.

There are several ways to resolve the issue of non-existence of an equilibrium. One option is to use weaker concepts of equilibrium such

[4]The exception is when the tie-breaking rule prescribes one of the servers with probability 1, while the price of the other server is already 0. However, in such a case the latter server can increase profits by a small price increase that will not affect the selection strategy of the customers.

as "ϵ-equilibrium" (see for example, §5.1). Another option is to modify the model by assuming that customers are heterogeneous or by making restrictive assumptions on the customers' behavior, as we now describe.

Li and Lee [107] considered two competing servers with observable queues and exponential service distributions with parameters μ_i, $i = 1, 2$. The joint arrival process is Poisson with rate λ. Customers are identical, with time value C and no balking allowed. They continuously monitor the queue lengths and have the option of jockeying, that is, instantaneously and costlessly moving from one queue to the (rear of the) other.[5] Server i sets a price p_i to maximize its average rate of profit. The payment is made upon completion of service to the server who completes the service.

We assume that $\mu_1, \mu_2 > \lambda$. The other cases are less interesting since with the no-balk assumption, if one of the servers is incapable of serving the whole population of customers, the other server can set an infinitely large price and gain an infinite profit rate by serving a positive fraction of the arrivals.[6]

As in the previous sections, the game has two stages. For a given value of $\delta = p_2 - p_1$, customers' equilibrium consists of a (possibly mixed) strategy that prescribes which of the servers to select for each state. Given these strategies, the servers' price equilibrium should be determined. As we observe below, the customers' equilibrium solution is not as simple as in the case where jockeying is not permitted. In particular, the server's selection of a new arrival does not depend solely on the difference between the queue lengths.

Similar to the model without jockeying, also here no equilibrium exists. To verify this assertion, suppose that an equilibrium does exist. Consider a customer who observes upon arrival i customers at server 1 and j customers at server 2. Since the customer prefers a lower payment, there exists a unique value $\delta = \delta_{ij}$ such that the customer is indifferent between the two queues, assuming that the others follow the equilibrium strategy. As explained earlier, such a value of δ cannot define an equilibrium since a server can increase profits, generally by an infinitesimal price reduction. If δ is different from δ_{ij} for all possible i and j, then again this is not an equilibrium as a server can increase profits by a small price increase.

Li and Lee concentrated therefore on a restricted model, assuming that:

[5] Jockeying is possible also for a customer currently in service.

[6] In the paper, Li and Lee assumed that there is an upper bound on the allowed price so that the other cases are also meaningful.

1 For some nonnegative integer m

$$p_1 + (m+1)\frac{C}{\mu_1} = p_2 + \frac{C}{\mu_2}.$$

2 The customers' strategy upon arrival when there are i customers at server 1 and j at server 2 is: join queue 1 if $i < j + m$; join queue 2 if $i > j + m$; randomly select each server with probability 0.5 if $i = j + m$. A customer jockeys similarly, with the exception that when he is indifferent between doing it or not, he does not jockey.

We note that rule 2 *is not a tie-breaking rule.* Indeed, when $i = j + m$ a customer is not indifferent between the two queues. To see this, consider a situation in which $\lambda \approx 0$ so that the possibility of a new arrival during the stay of the current arrival in the system is negligible. Assume further that $i = m$ and $j = 0$. In this case, the selection rule prescribes a random choice. However, joining queue 2 is clearly a better choice since the customer immediately starts service, without giving up the option of jockeying to the faster server if that server becomes idle while the customer is still in the system. Similarly, the assumed customers' strategy may not be a best response in other states. Therefore, the model is about servers' equilibrium but not customers' equilibrium.

Li and Lee justify this model as follows:

> Though the restriction is artificial, the model still captures the essential trade-offs in time competition. Also, we always have in mind a continuous approximation of the restricted market shares which retains the essential properties of time competition.

Consequently, using n to denote the total number of customers in the system:

- If $n > m + 1$, both servers are busy.
- If $n < m + 1$, server 1 is busy and server 2 is idle.
- If $n = m + 1$, either all customers are in queue 1, or m are in queue 1 and one is in queue 2.

The resulting system becomes essentially one dimensional with the state variable n, except for when $n = m + 1$ which corresponds to two possibilities. The latter case has to be "broken" into two states to get a Markov process with this state space. Consequently, it is possible to obtain explicit formulas for the proportion, $\gamma(m)$, of the customers whose service is completed by server 1.

In equilibrium, server 1 maximizes $p_1\gamma(m)$ over p_1, taking p_2 as given, and server 2 maximizes $p_2[1-\gamma(m)]$ over p_2, taking p_1 as fixed, both under the constraint $m = \left[\frac{p_2-p_1}{C} + \frac{1}{\mu_2}\right]\mu_1 - 1$. Note that if, for example, server 1 selects a very high price then the roles of 1 and 2 may interchange according to the definition.

THEOREM 7.2 *(Li and Lee [107]) There are numbers $\underline{p}_1$ and $\bar{p}_1$ such that (p_1, p_2) is an equilibrium if and only if $\underline{p}_1 \le p_1 \le \bar{p}_1$ and $p_1 + \frac{C}{\mu_1} = p_2 + \frac{C}{\mu_2}$.*[7]

Thus, in equilibrium, $m = 0$. Another result that follows from the theorem is that

- the faster server sets a higher price and still gets a higher share of the market.

4. Price and priority competition

Lederer and Li [101] considered a competitive multi-server extension of the model of Mendelson and Whang [124] (§4.4.3). In their model, customers belong to *classes.* Service of customers of class z ("z-customers"), is done at a rate of μ_z, and their time value is C_z. Servers can discriminate among customers by their class. Server i selects a priority rule f_i, and charges z-customers a price $p_i(z)$. Denote by $\underline{\lambda}_i = (\lambda_i(z))$ the vector of arrival rates to queue i whose priority rule is f_i. A z-customer selects a server to minimize his expected full price $P_i(z) = p_i(z) + C_z W_i(z, \underline{\lambda}_i, f_i)$, where $W_i(z, \underline{\lambda}_i, f_i)$ is the expected waiting time of a z-customer who joins queue i. Denote the server's cost when serving the arrival rates of $\underline{\lambda}_i$ by $c_i(\underline{\lambda}_i)$.

Lederer and Li assumed that the number of servers is large so that each server's influence on the equilibrium full prices is negligible. They also assumed that entry of new servers is not possible (a short-term model).

In equilibrium, the servers with $\lambda_i(z) > 0$ have the same expected full price P_z for z-customers:

$$P_z = p_i(z) + C_z W_i(z, \underline{\lambda}_i, f_i).$$

Therefore, the profit that server i gets from serving z-customers can be written as

$$\lambda_i(z)p_i(z) = \lambda_i(z)[P_z - C_z W_i(z, \underline{\lambda}_i, f_i)].$$

[7] The two conditions also imply lower and upper bounds on p_2.

This equation also holds when $\lambda_i(z) = 0$, since in this case both its sides are 0. Hence, the total profit of server i is

$$\Pi_i(\underline{P}, \underline{\lambda}_i, f_i) = \sum_z \lambda_i(z)[P_z - C_z W_i(z, \underline{\lambda}_i, f_i)] - c_i(\underline{\lambda}_i), \tag{7.5}$$

where $\underline{P} = (P_z)$.

Following [124], Lederer and Li considered the server's problem as that of maximizing $\Pi_i(\underline{P}, \underline{\lambda}_i, f_i)$ by selecting a vector of arrival rates $\underline{\lambda}_i$ and queue disciplines f_i.

Applying Little's formula to (7.5) we obtain that for server i, $i = 1, \ldots, n$,

$$\Pi_i(\underline{P}, \underline{\lambda}_i, f_i) = \sum_z [P_z \lambda_i(z) - C_z L_i(z, \underline{\lambda}_i, f_i)] - c_i(\underline{\lambda}_i), \tag{7.6}$$

where $L_i(z, \underline{\lambda}_i, f_i)$ is the expected number of z-customers at the i-th queue.

For any given $\underline{\lambda}_i$, (7.6) is maximized when the service discipline minimizes $\sum_z C_z L_i(z, \underline{\lambda}_i, f_i)$. It is well known that under fairly general conditions, the minimizing discipline is that which gives higher priority to customer classes with higher $C_z \mu_z$ values (the $C\mu$-rule) (see [52]). Therefore, this rule will be adopted by all servers. The variable f_i can now be omitted, and the i-th server's problem is to choose $\underline{\lambda}_i$ to maximize $\Pi_i(\underline{P}, \underline{\lambda}_i)$ as in (7.5), for any given $\underline{P}$.

Suppose that class z has a demand function d so that given the full price P_z, the arrival process of z-customers is Poisson with rate $d(z, P)$. An equilibrium is a set of vectors $\underline{P}$ and $\underline{\lambda}_i$ for every server i, such that $\underline{\lambda}_i$ solves server i's profit-maximizing problem, given $\underline{P}$ and $d(z, P) = \sum_i \lambda_i(z)$ for every class z.

Lederer and Li proved the following results:

- If $\sum_z C_z L_i(z, \underline{\lambda}_i) + c_i(\underline{\lambda}_i)$ is strictly convex in $\underline{\lambda}_i$ for every i, then an equilibrium exists.
- Equilibria are incentive-compatible.
- If there is just one class of customers then there exists a unique equilibrium.
- Suppose that the service requirements of all classes have an identical exponential distribution, and that for every i, $c_i(\underline{\lambda}_i) = c_i[\sum_z \lambda_i(z)]$. Then there exists a unique equilibrium.

5. Search among competing servers

Davidson's [40] model is a combination of the generic models of observable and unobservable queues. Davidson considered a large number of servers so that the queue lengths at these servers are assumed to be independent. Customers have heterogeneous time values and every customer knows his own value. Balking is not allowed, and customers search among the servers until they decide to join.

The arrival rate is λ *per server*. The cost incurred by a customer with time value C (a "C-customer") due to an inspection of a queue is $b + wC$, where b and w are constant across customers (w is interpreted as the search time). The goal of a customer is to minimize his expected costs due to search, admission, and waiting.

A C-customer's strategy is defined by a threshold $B(C)$ on his full price. He joins a server who charges p and whose queue size is n if and only if his expected full price is at most $B(C)$, that is, if and only if $p + C\frac{n+1}{\mu} \leq B(C)$.

- **Incomplete information** Assume that customers are uninformed on the prices charged by the servers. Thus, by visiting a server the customer learns both its price and its queue length.

  Since customers do not know the prices in advance, the arrival rate to a server is independent of its price. After learning the price and observing the queue, an arriving customer decides whether to join or to continue searching. Being aware of the customers' decision process, each server selects a price that optimizes its expected profits. The arrival rate to a server is independent of its price, and we are looking for a symmetric equilibrium under which all servers select the same price.

  Davidson didn't specify how to compute the equilibrium threshold function $B(C)$. We now briefly outline this process. Note that since balking is not allowed in this model, the customers only optimize their search and waiting costs, and the equilibrium search strategies are independent of the prices. If a price p is required by all servers, this is a fixed cost that every customer must pay, and the equilibrium threshold full price will be increased by p. Let $B_0(C)$ be the equilibrium threshold function, given that $p = 0$. Let q_i denote the steady state probability of i customers at a random server, when $B_0(C)$ is adopted by all customers. Let $A_C(\beta)$ denote the expected cost incurred by a C-customer who chooses a threshold β, given that everybody else follows $B_0(C)$. Let $i(\beta, C) = \max\{i : C(i+1) \leq \beta\mu\}$ denote the largest queue size such that a C-customer with threshold

β still joins, given that the admission fee is 0. Then,

$$A_C(\beta) = b + wC + \sum_{i \le i(\beta,C)} q_i C \frac{i+1}{\mu} + A_C(\beta) Pr[i > i(\beta, C)],$$

or

$$A_C(\beta) = \frac{b + wC + \sum_{i \le i(\beta,C)} q_i C \frac{i+1}{\mu}}{Pr[i \le i(\beta, C)]}.$$

One should find from this relation the best response, and then the equilibrium function $B_0(C)$. In the next stage, the equilibrium price can be found by

$$p_e = \arg\max_p \{p \int_C Pr[i \le i(B_0(C) + p_e - p, C)]dC\}.$$

- **Complete information** In this case the prices are advertised by the servers, and customers search among them to reveal their queue lengths. Davidson considered a simplified model in which a proportion α of the customers has $C = 0$, and the rest have $C = 1$. Note that the assumption that one of the cost values is 0, is part of the model, whereas the assumption that the other cost value is 1, is without loss of generality. The respective thresholds are denoted by B_0 and B_1. 0-customers are indifferent to the queue length, and hence join only those servers that offer the lowest price, p_0. 1-customers may join servers with different prices, as long as the expected full price at these servers is the same. Davidson concluded that: "Since consumers do not care which type of server they end up at, it seems logical to assume that they will choose the server randomly, and all servers will face the same arrival rate." This conclusion is problematic since the fact that customers are indifferent in regard to a set of servers, does not mean that their arrival rates are arbitrary (see Remark 1.1). The rates should be set at those levels which make their expected full prices equal. It is an open problem to continue Davidson's research. Another seemingly harder open problem is a model that assumes a small number of servers so that the queue lengths are not independent.

6. Information based competition

Hassin [68] raised the question of whether a server can profit by concealing the information on the length of the queue. We next describe Hassin's model.

There are two servers with separate queues denoted by Q1 and Q2. The servers are identical except that the length of Q1 can be observed, whereas the length of Q2 cannot be observed.[8] The input to the system consists of a Poisson stream of customers with rate λ. Each customer wishes to minimize his expected waiting time. The service is exponentially distributed with rate μ. Upon arrival, a customer observes Q1, and decides which of the two queues to join. The decision is irrevocable. An arriving customer bases his choice on the actual state at Q1, and on the expected length of Q2 *conditioned on the length of Q1*. Clearly, this conditional expectation is a function of the choice strategies of the others customers.

Hassin investigated threshold strategies (§1.2). As we show below, the individual's best response threshold is monotone non-increasing with the threshold of the others in the population: the higher their tendency to enter Q1, the lower the tendency for the individual to do so. Thus, this is an ATC model with a unique equilibrium threshold strategy.

For a positive number x, let $n = \lfloor x \rfloor$ and let $r = x - n$.[9] Under a threshold strategy with threshold x, a customer joins Q1 if its size is at most $n-1$, joins Q2 if Q1 has more than n customers, and selects Q1 with probability r (and Q2 with probability $1-r$) when Q1 has exactly n customers.

Note that under a pure strategy ($r = 0$) with a threshold n, a customer enters Q2 whenever he observes n customers in Q1. Thus, in this case, n is the maximum length of Q1. In general, if all the customers in the population follow a threshold strategy with threshold x then the maximum possible number of customers in Q1 is $\lceil x \rceil$.

6.1. Existence of an equilibrium

Let $f(L, x)$ denote the expected length of Q2, given that the length of Q1 is L, and everyone follows the threshold strategy x. Let $g(L, x) = f(L, x) - L$ denote the expected difference between the lengths of Q2 and Q1, given the length L of Q1 and that the threshold x is used by all. Hassin constructed the function g for various values of the utilization factor $\rho = \frac{\lambda}{2\mu}$ and found that it is monotone decreasing both in L and in x.

Consider a customer who assumes that others follow the threshold strategy x. His best response is to enter Q1 if $g(L, x) > 0$, and to enter Q2 if $g(L, x) < 0$. He is indifferent between the two options if

[8]Hassin illustrated the model by considering two gas stations located one after the other on a main road. A driver who needs to fill his tank sees the queue at the first station, but not at the second one.

[9]$\lfloor x \rfloor$ is the largest integer which is less than or equal to x. Therefore, $0 \le r < 1$. Similarly, $\lceil x \rceil$ is the smallest integer which is greater than or equal to x.

$g(L,x) = 0$. By the monotonicity of $g(L,x)$ with respect to L, it follows that for a given value x, there is a maximal size of L such that entering Q1 is a best action.

Specifically, the customer's best response is of the threshold type: let $k(x) = \min\{L : g(L,x) \leq 0\}$; if $L \in \{0,\ldots,k(x)-1\}$ join Q1, otherwise join Q2. By the monotonicity of $g(L,x)$ with respect to x, it follows that $k(x)$ is a non-increasing step function. Hence, this is an ATC situation and, in particular, a unique equilibrium exists.[10] The discontinuity points of $k(x)$ correspond to thresholds x such that the individual is indifferent between two consecutive thresholds.

Denote by p_{ij} the steady state probability of i customers at Q1 and j customers at Q2, when all use the threshold strategy x. For simplicity, we suppress the reference to x in this notation.

Consider first equilibria in pure strategies. In order for the integer n to describe a best response for an individual, given that all follow this strategy, two conditions are necessary: (1) if a customer arrives and observes $n-1$ customers in Q1 his optimal choice is to join this queue, and (2) if he observes n customers, his optimal choice is to join Q2. Moreover, by the monotonicity of $g(L,x)$ in L, these conditions are also sufficient: if it is optimal to join Q1 when $n-1$ customers are observed, then it is also optimal to do so when fewer customers are observed; similarly, if it is optimal to join Q2 when n are observed, then this is also the optimal choice when more than n are observed.

Let p_{ij} be the steady-state joint probability of having i customers in Q1 and j customers in Q2. Let $q_{j|i} = \frac{p_{ij}}{\sum_{k=0}^{\infty} p_{ik}}$ be the probability that the length of Q2 is j given that the length of Q1 is i. To simplify notation we have suppressed the reference to x in these probabilities, but clearly they are function of the strategy adopted by all. Thus, n describes an equilibrium if and only if

$$\sum_{j=0}^{\infty} j q_{j|n} \leq n \leq 1 + \sum_{j=0}^{\infty} j q_{j|n-1}.$$

In particular, the threshold $n = 1$ is an equilibrium if and only if $\sum_{j=0}^{\infty} j q_{j|1} \leq 1$.

Consider now conditions for an equilibrium in mixed strategies. In order for the threshold $x = n + r$, $0 < r < 1$, to define an equilibrium when all follow it, an arrival who observes n customers in Q1 must be

[10] Yechiali, Altman, Jimenez, and Núñez-Queija [180] showed that this result does not hold when the servers have different service rates, and it may be then that there is no equilibrium threshold strategy.

indifferent between joining Q1 or Q2. Thus, the threshold $x = n + r$, $0 < r < 1$, specifies an equilibrium if and only if

$$n = \sum_{j=0}^{\infty} j q_{j|n}.$$

6.2. Solution of the model

Let $\mathbf{I}$ denote the indicator function, so that $\mathbf{I}(R) = 1$ if the relation R is satisfied, and $\mathbf{I}(R) = 0$ otherwise. Let $x = n + r$ be as before. Then, the steady state probabilities satisfy the following equations:

$$\begin{aligned}
\lambda p_{ij} + \mu p_{ij}\mathbf{I}(i > 0) + \mu p_{ij}\mathbf{I}(j > 0) &= \mu p_{i,j+1} \\
&+ \mu p_{i+1,j}\mathbf{I}(i \le n-1, \text{ or } i = n \text{ and } r > 0) \\
&+ \lambda p_{i-1,j}\mathbf{I}(0 < i < n) \\
&+ \lambda r p_{i-1,j}\mathbf{I}(i = n+1 \text{ and } r > 0) \\
&+ (1-r)\lambda p_{i,j-1}\mathbf{I}(i = n \text{ and } j > 0) \\
&+ \lambda p_{i,j-1}\mathbf{I}(i = n+1 \text{ and } j > 0).
\end{aligned}$$

Hassin solved numerically for the equilibrium x and found the following results:

- x is a nondecreasing function of ρ. This can be explained in view of Figure 1.2: a change in ρ shifts the function $k(x)$ and the intersection with the identity function is obtained at the same integer value in a non-degenerate interval of ρ.
- Let λ_i denote the equilibrium arrival rate to server i, $i = 1, 2$. Then, $\frac{\lambda_1}{\lambda_2}$ is not a monotone function of ρ, though its general trend is decreasing in ρ.
- $\frac{\lambda_1}{\lambda_2} > 1$ for all values of ρ, so that the equilibrium arrival rate to Q1 is larger than that of Q2: the server that reveals the queue length information gets a higher share of the demand.
- Clearly, if both queues can be observed the demand is split equally between the two servers. Thus, the server of Q2 has an incentive to supply the queue size information to the customers. The demand is also split equally if none of the two queues can be observed. However, this situation is not an equilibrium with respect to the servers' behavior: each of them has an incentive to reveal the size of his queue and increase the fraction of the demand directed to his facility.[11] We thus

[11]The situation corresponds to a two-person constant-sum game, where each server chooses between two strategies: to reveal or not its queue lengths. In this game, the strategy of revealing the queue length is a dominant strategy.

expect to find observable queues whenever this is technically possible and costless.[12]

7. Related literature

- Chen and Wan [35] considered an unobservable model similar to the one described in Sections 1 and 2, but assumed that the servers may differ in their service rates. They obtained the following results:

  - There may be a unique equilibrium, no equilibrium, or a continuum of equilibria.
  - In the case of a unique equilibrium, three cases are possible: one server takes the whole market; both servers charge their monopoly optimal prices; both servers charge non-monopolistic prices that leave positive consumer surplus.
  - If the service rates of the two servers are equal, an equilibrium always exists. It may be unique or there may be a continuum of equilibria.

- Tapiero and Zuckerman [167] presented a model of competition between two transport servers. Server i operates a vehicle of capacity M_i. The vehicle is dispatched after T_i time units or when the number of waiting customers is M_i, whichever occurs first. The customers' arrival process is Poisson with rate λ. Thus, customers are not aware of the schedule.[13] The probability that a customer joins queue i is a function of the prices p_1, p_2 and the expected waiting times W_1, W_2. Unlike most of the other models in this chapter, Tapiero and Zuckerman did not assume that the full price in equilibrium is identical at the two servers.

  The decision variables are the price p_i, the capacity M_i, and the dispatching interval T_i. Attention is given to special cases in which either $T_i = \infty$ or $M_i = \infty$.

  There are no closed-form formulas for the equilibrium solution, and further research is required to reach a better understanding of the model.

- Other models of competition, which also involve the long-run decision of determining the service capacity, are described in §8.

---

[12] In the application to the case of two gas stations located on the same road, the first one has a higher profit. A natural outcome is that the second station tries to attract demand by reducing prices or offering other benefits. Equilibrium prices can then be computed. This is an open problem.

[13] See §6.3 for a model in which customers react to the schedule.

Chapter 8

SERVICE RATE DECISIONS

This chapter is concerned with models in which the service rate is a decision variable. In most cases, it is the server who determines the service rate, but we also deal with models where this is done by the customers. We also consider in this chapter models where the customer determines the amount of service he obtains.

The problem of determining optimal prices to regulate the arrival process to a queueing system is considered by several authors as a *short-run problem.* In contrast, in the *long-run problem*, the facility manager also controls the service capacity. The long-run problem is considered in this chapter.

As shown by Edelson and Hildebrand (§3.1), in an unobservable system with homogeneous customers, the social and profit-maximizing objectives are identical. Consequently, the service rate chosen by a profit maximizing server in the long-term problem is also socially optimal. This is not necessarily true in other models, as will be discussed below.[1]

We denote by $c(\mu)$ the cost per unit of time associated with operating service at the rate of μ. In most models it is assumed that this operating cost is independent of the utilization of the server.

All the models described in this chapter assume unobservable queues, with the exception of Section 5 where the basic observable and unobservable models are compared, under the assumption that the cost of operating a server is independent of its utilization.

[1]The service rate can be viewed as a quality parameter for the product supplied by the firm. See Spence [159] for a general discussion on the profit-maximizing price and quality set by a monopoly.

1. Heterogeneous service values

We first describe the results of Mendelson [123] for the long-run version of the unobservable queueing model given in §3.3. The social planner's problem in this model is

$$\max_{\lambda,\mu\geq 0}\{V(\lambda)-CL(\lambda,\mu)-c(\mu)\}, \tag{8.1}$$

where L, the expected number of customers in the system, has the form $L(\lambda,\mu) = f(\rho)$ for some function f, where $\rho = \frac{\lambda}{\mu}$. For example, in the $M/M/1$ model $f(\rho) = \frac{\rho}{1-\rho}$. A natural assumption is that $c(\mu)$ is monotone increasing and convex, whereas $V(\lambda)$ is monotone increasing and concave. It is assumed further that $V'(0)$ is sufficiently large, since otherwise $\lambda = \mu = 0$ would have been the optimal solution. Moreover, $\lim_{\lambda\to\infty} V'(\lambda) \leq \lim_{\mu\to\infty} c'(\mu)$. This assumption guarantees that the objective function in (8.1) is bounded.

The first-order conditions for social optimality are

$$V'(\lambda) = \frac{C}{\mu}f'(\rho) \tag{8.2}$$

and

$$\frac{C\lambda}{\mu^2}f'(\rho) = c'(\mu). \tag{8.3}$$

These conditions imply that

$$\lambda V'(\lambda) = \mu c'(\mu). \tag{8.4}$$

An admission fee p induces an equilibrium arrival rate λ which equates the marginal value $V'(\lambda)$ and the expected full price $p + CW(\lambda,\mu)$. Hence, given the optimal service rate μ^*, the optimal arrival rate λ^* can be induced by an admission fee p^* such that

$$V'(\lambda^*) = p^* + CW(\lambda^*,\mu^*). \tag{8.5}$$

Assume that $c(\mu) = a + b\mu$ for some $a, b > 0$, then by (8.4) and (8.5),

$$\lambda^* p^* = \lambda^* V'(\lambda^*) - \lambda^* CW(\lambda^*,\mu^*) = b\mu^* - \lambda^* CW(\lambda^*,\mu^*).$$

Thus, the profit is equal to the variable production costs minus the waiting costs:

- With linear operating costs, and under a social welfare maximizing fee, the service facility fully bears the waiting expenses of its customers.

Unlike the short-run case, comparison to the profit-maximizing solution does not yield conclusive results. Mendelson concludes that the outcome depends on the model's parameters. In particular:

- In some cases the profit maximizer selects a solution in which the utilization factor ρ is smaller than the socially optimal value. In such cases, the quality of service supplied by the monopoly, as reflected by ρ, is higher than the socially optimal one, since this allows an increase in price and profits (while decreasing the number of customers who enjoy service).

Finally, Mendelson demonstrated through an example that

- A low utilization factor is not an indication for inefficiency, but it may be a consequence of the need to shorten the waiting time.

Dewan and Mendelson [44] found the following interesting relationship.[2] Assume an $M/M/1$ model with linear operating costs, $c(\mu) = a + b\mu$. Then, substituting $f(\rho) = \frac{\rho}{1-\rho}$, (8.2) and (8.3) give, respectively,

$$V'(\lambda) = \frac{C\mu}{(\mu-\lambda)^2} = \frac{C}{\mu-\lambda} + \frac{C\lambda}{(\mu-\lambda)^2},$$

and

$$b = \frac{C\lambda}{(\mu-\lambda)^2}.$$

Together, these equations imply that

$$V'(\lambda) = \frac{C}{\mu-\lambda} + b$$

which coupled with (8.5) lead to

$$p^* = b.$$

In words:

- In an $M/M/1$ system with linear operating costs, the optimal admission price equals the marginal cost of increasing the service rate.

This result also extends to a model with several distinguishable customer classes, each having its own utility function and (nonlinear) waiting cost function. Yet, the service requirement has to be exponential with a common rate to all classes.

[2] A similar result was established by Balachandran and Radhakrishnan [19].

Mendelson's model was extended by Stidham [165] in several aspects. Stidham showed that the long-run problem typically has multiple solutions. This may happen even in an $M/M/1$ model with linear delay costs.[3]

2. Service rate at a fixed price

Stenbacka and Tombak [162] considered a model of an unobservable queue in which an admission fee p is exogenously regulated and cannot be altered by the server. The server can maximizes profits by varying the service rate. Customers differ by their time values C that are distributed according to a differentiable distribution function $F(C)$ with density $f(C)$. All customers value service by R. Customers have the option of balking and the effective arrival rate is therefore affected by the expected waiting time. Stenbacka and Tombak compared the socially optimal solution with the profit-maximizing one. They gave conditions under which the monopoly provides worse service (i.e., with a rate slower than the optimal rate) and discussed the effect of these findings on privatization of public services.

The model of Stenbacka and Tombak is more general than the one described below. We modify their model by assuming an $M/M/1$ queue. We then get simpler and more intuitive sufficient conditions guaranteeing their results. Recall that $c(\mu)$ denotes the cost (per unit of time) of operating service at rate μ, and let $W(\mu)$ denote the expected waiting time when the service rate is μ. In our case $W(\mu) = \frac{1}{\mu-\lambda}$.

A C-customer chooses to join the queue if $p + CW(\mu) \leq R$. Thus, only customers whose time value C is at most some threshold value C_e join, where C_e is determined by[4]

$$C_e = \frac{R-p}{W(\mu)} = (R-p)[\mu - \lambda F(C_e)].$$

Note that C_e is a function of μ but to simplify notation we use C_e instead of $C_e(\mu)$. We also denote the derivative of C_e with respect to μ by C_e'.

Social welfare is

$$S(\mu) = \lambda \int_{C \leq C_e} \left(R - \frac{C}{\mu - \lambda F(C_e)} \right) f(C)\, dC - c(\mu).$$

---

[3]Recall that in the short-run problem, where μ is fixed, there is a unique equilibrium under these conditions. See Figure 3.4.

[4]We deviate here from the model of Stenbacka and Tombak in which it is implicitly assumed that the waiting time is independent of the arrival rate. See Equation (4) and the definition of the threshold in [162].

The first-order condition for social optimality is

$$\frac{dS(\mu)}{d\mu} = \lambda p f(C_e)C_e' + \lambda \int_{C \le C_e} \frac{\lambda C f(C_e)C_e'}{[\mu - \lambda F(C_e)]^2} f(C)\, dC - c'(\mu) = 0. \quad (8.6)$$

The server's profit is

$$Z(\mu) = \lambda p F(C_e) - c(\mu).$$

The first-order condition for maximizing profit is

$$\frac{dZ(\mu)}{d\mu} = \lambda p f(C_e)C_e' - c'(\mu) = 0. \quad (8.7)$$

The two conditions (8.6) and (8.7) differ by the nonnegative term

$$\int_{C \le C_e} \frac{\lambda C f(C_e)C_e'}{[\mu - \lambda F(C_e)]^2} f(C)\, dC.$$

This term is nonnegative since $C_e' \ge 0$. Thus, under the socially optimal service rate the term $\lambda p f(C_e)C_e' - c'(\mu)$ is smaller than the corresponding value under the profit maximizing service rate. If the distribution function F and the operation cost function c are such that this term is monotone increasing in μ, then the service rate chosen by the monopoly is smaller than the socially optimal one. However, this property does not necessarily hold in general.

Stenbacka and Tombak considered another model of unobservable queues which they call *time-based competition*, where again, prices are fixed and the decision variable is the service rate. They compared the waiting time obtained when there is one private and one public server, possibly with different cost functions, to that obtained when the public firm is privatized. They show that privatization often decreases the equilibrium service rates. It should be emphasized that the model assumes that prices do not change as a result of the change of ownership. Also, the authors assume that payments to a private firm are real costs and not merely transfer payments (as are payments to the public firm).

3. Bribes and auctions

Several papers ask whether bribes and auctions (§4.5) may cause the server to slow down service so that customers will be induced to offer higher payments. Myrdal [131] claimed that corrupt officials may deliberately cause administrative delays so as to attract more bribes. Lui [112] referred to this claim as *Myrdal's hypothesis*, and argued that the hypothesis is not always true. For example, if increasing the rate of service is costly to the server, then without a bribe the server has no incentive to supply service, and bribes induce faster service. In contrast to this point of view, Hassin [67] compared the service rate chosen by a profit maximizer to the socially optimal rate, showing that from this point of view Myrdal's hypothesis is correct:

THEOREM 8.1 *In an unobservable $GI/M/1$ system in which priorities are determined by customers' payments, the service rate chosen by a profit maximizing server is smaller than or equal to the socially optimal rate.*

Hassin proved the theorem also when customers are heterogeneous in their service values and time costs. We will only describe the simpler case of homogeneous customers.

If the server is slow, then in equilibrium, a fraction of the potential demand will choose not to join the queue. If, on the other hand, the server is fast, then all of the potential demand (with rate Λ) is served. Thus, there exists a cutoff value, μ_0, such that $\lambda = \Lambda$ if and only if the service rate is at least μ_0.

In this section, we denote by Z and S the profit and social welfare, *excluding costs associated with operating the service.*

Consider first the case where $\mu < \mu_0$. An increase in μ has two opposing effects on the average waiting time: each customer's service is made shorter, but more customers are attracted to join the queue. However, since customer behavior is socially optimal (§4.5), social welfare increases when the service is made faster (it will increase even if λ remains unchanged, all the more so when it changes to its new optimal value). In this range, $Z = S$ (see (4.22)). Therefore, for $\mu < \mu_0$

$$\frac{dZ}{d\mu} = \frac{dS}{d\mu} > 0.$$

Consider now the case where $\mu > \mu_0$. An increase in μ will not affect the arrival rate, which already consists of the whole potential demand. Thus, average waiting time must decrease, thereby increasing social welfare. By (4.21) the server's profit decreases, because faster service reduces the difference between the expected waiting time of a low priority customer and that of a random customer. Thus the same number of customers is served but customers pay less. We conclude that for $\mu > \mu_0$

$$\frac{dZ}{d\mu} < 0, \quad \frac{dS}{d\mu} > 0.$$

When μ approaches infinity the expected waiting time, even of the customer with the lowest priority, as well as the expected service time, decrease to 0. Therefore,

$$\lim_{\mu \to \infty} S = \Lambda R, \qquad \lim_{\mu \to \infty} Z = 0.$$

Figure 8.1 illustrates these conclusions. We observe that if the service rate can be controlled costlessly, then μ_0 is the rate chosen by the profit maximizer; all of the potential demand arrives to the queue, and customer's welfare is 0. Clearly, the socially optimal rate is infinite.

Recall that operating a facility with a service rate of μ involves a cost of $c(\mu)$ per unit time. The following two cases are possible:

- The socially optimal rate is larger than μ_0. For example, when $c(\mu)$ is represented by the function c_1 in Figure 8.1, the optimal rate is μ_1. In this case the profit maximizing server chooses a slower service rate (frequently this will be exactly μ_0, as in Figure 8.1).

- The socially optimal rate is smaller than or equal to μ_0. For example, when $c(\mu)$ is given by the function c_2 in Figure 8.1. In this case the server voluntarily chooses this optimal rate (μ_2 in Figure 8.1), because it also maximizes profits.

4. Asymmetric information

This section describes long-run decisions in models where the customers are heterogeneous, having different time or service values, but this private information is not available to the server who cannot use it to discriminate among customers.

4.1. Heterogeneous service values

The model of Whang [173] resembles the unobservable model of §3.3, except that the queue manager and the customers do not know the marginal value function $V'(\lambda)$. Thus, the service value of a customer is his private information.

Whang modeled the situation as a non-cooperative game. The game proceeds in several stages:

- Customers anonymously report their service values. The meaning of anonymity is that these reports cannot be used to discriminate among customers. Denote the resulting reported marginal value function by $m(\lambda)$.

- Using these reports, the server makes the long-run decision of choosing a capacity $\mu(m)$. This comes at a linear cost $c(\mu) = b\mu$.

- The actual function V' is realized and observed and the server uses it for the short-run decision of the service fee $p(V', \mu)$. It is assumed that the server can commit to fixed rules $\mu(m)$ and $p(V', \mu)$.

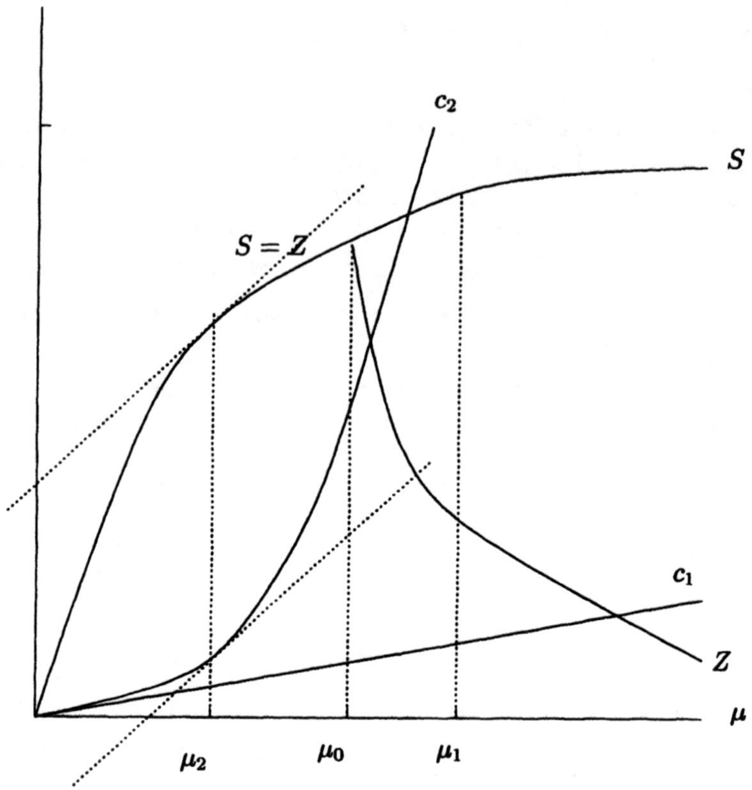

Figure 8.1. Optimal service rates in a system with bribery

- Lastly, the equilibrium arrival rate, given μ and p, is obtained, as in §3.3.

Whang observed that those customers who choose to arrive have the same objective with respect to the server's long- and short-run decisions, namely, they all wish to minimize $CW(\lambda, \mu)+p$. Therefore, it is assumed that customers cooperate in their reports so that a customer with service value R reports a value $\sigma(R)$ so that the resulting rate, m, solves

$$\min_m [CW(\lambda, \mu(m)) + p(V', \mu(m))]$$

subject to the equilibrium condition

$$V'(\lambda) = CW(\lambda, \mu(m)) + p(V', \mu(m)).$$

The rule σ may depend on the customers' service value, but not on the unknown function V'.

Whang observed that if according to the rule $\mu(m)$ higher reports lead to a higher service rate, then customers will report increasingly high values so that a higher than social optimal μ is set by the server. Therefore, an optimal rule for the server is not monotone.

A strategy $(\mu(m), p(V', \mu))$ that achieves a solution which is socially optimal under full information is said to be a *full-information-efficient strategy.* Whang's main claim is that if $W = W(\rho)$, then the following strategy is full-information-efficient and motivates each customer to reveal his true service value:

- On receiving the reports m, the manager should solve the system's problem assuming that the reports are true, and fully allocate the capacity cost $b\mu^*(m)$ to the customers.

Note that the assumption $W = W(\rho)$ does not hold in most queueing models, in particular $M/M/1$.

4.2. Heterogeneous time values

Balachandran and Radhakrishnan [19] analyzed a model of class decision with asymmetric information. In this model, the demand for service consists of a finite number of classes of customers each of which with *given rates of demand.* The cost of operating the server is a convex monotone increasing function $c(\mu)$. The time values, C_i, are private information of the controllers of the classes. The sequence of events is as follows:

- The server announces a rule by which the server's capacity is determined and the operating costs are divided among the classes.
- Classes report their time values, C_i^r (that may be different from the true values, C_i).
- The server decides on the service rate μ.
- Classes pay for the operating costs according to the rule.

Balachandran and Radhakrishnan observed that if the server determines the overall optimal μ assuming that the time values are as reported, then classes have incentives to overstate their time values and by doing so induce higher service rates and shorter waiting costs. It is assumed therefore that the server can obtain measures, X_i, of the true values C_i. These measures are required to be unbiased estimators for C_i (that is, $E(X_i) = C_i$). Balachandran and Radhakrishnan show that such measures make it possible to determine a rule for allocating the operating costs so that classes are induced to report their true time values.

It should be emphasized that the purpose of the pricing scheme here is not to control the arrival rates (these are considered to be fixed) but to induce truthful reports of the time costs in order to determine optimal service capacity.

5. Observable vs. unobservable queues

Hassin [66] compared the social welfare under the profit-maximizing number of single server facilities, in the basic models of Naor [133] and Edelson and Hildebrand [47]. Let c denote the cost per unit of time associated with operating a facility, regardless of its utilization. Assume that the potential arrival rate is large and that the queue organizer determines the optimal arrival rate to each facility. If the gain from a facility that serves an arrival process with rate λ is $Z(\lambda)$, then the optimal arrival rate per facility maximizes $\frac{Z(\lambda)-c}{\lambda}$, that is, maximizes gain per unit of arrival rate.

Suppose that the service manager can choose between revealing the queue length to the customers and operating an observable queue, or concealing this information and operating an unobservable queue (§3.2). Hassin showed that social welfare may, in some cases, be increased by motivating the profit maximizing firm to reveal the queue length when it otherwise prefers to conceal it. On the other hand, it never pays, from the social point of view, to induce the firm to conceal the queue length when it is willing to reveal it. To see this let the operating cost c be such that the firm prefers to reveal the queue length. Let λ_1 be the rate it chooses in this case, and let λ_2 be the rate it would choose had it been impossible for it to reveal this information. Then,

$$\frac{S_O - c}{\lambda_1} \geq \frac{Z_O - c}{\lambda_1} > \frac{Z_U - c}{\lambda_2} = \frac{S_U - c}{\lambda_2},$$

where Z denotes profit, S denotes social welfare, O denotes observable queues and U denotes unobservable queues. The first inequality holds since $S_O \geq Z_O$, the second since the firm prefers an observable queue. Thus, social welfare decreases when the firm operates an unobservable queue at rate of λ_2 .

6. Co-production

Ha [62, 63] considered an unobservable system where each customer chooses a service rate. The choice reflects the *amount of service* he requests, and affects his utility from service. Cachon and Harker [31] (see Section 7) made a similar assumption in their model. They refer to it as an engagement of customers in *co-production* or *outsourcing* of service to customers. Note that a high rate of service means that less service is given by the server, whereas longer service is associated with higher value to the customer.

The decrease in utility that a customer obtains from faster service is reflected in a cost function $h(\mu)$ incurred by a customer if he chooses service rate μ. The function $h(\mu)$ is continuously differentiable, monotone increasing, and strictly convex.

Customers ignore the externalities of their choice of a service rate on the delays incurred by others and choose a rate which is smaller than the service rate a social planner would choose.[5] Ha derived pricing schemes under which customers make decisions that are compatible with social optimality.

Ha's approach distinguishes between two types of externalities involved with the decision process of co-production models. *Service externalities* are associated with the individual's optimization of his service requirement. Specifically, the longer a customer's service, the more waiting time is added to others. When choosing a service rate, the customer ignores these externalities and therefore his choice tends to be too small from a social point of view. *Admission externalities* are involved with the increase in arrival rate caused when more customers decide to join. We note that this distinction is mainly semantic: The two types of externalities are caused by the same action, that of joining the queue in order to obtain service.

6.1. Single class FCFS model

The first model of Ha [62] assumes a $GI/GI/1$ FCFS queue with customers who are identical in all parameters except for their willingness to join the system at a given value of the expected full price. Note that the full price in this model includes the cost $h(\mu)$ (see (8.11)).

[5]The following illustrative situation is described by Schelling [153]. An accident occurs in an freeway. Drivers in the opposite lane slow down to watch, creating long lines of cars behind them. Eventually, many commuters spend ten minutes extra driving for a ten-second look. When they get to the scene, the ten minutes' delay is a sunk cost, and they pay the extra ten seconds for their own sightseeing. As a collective body, the drivers might vote to maintain speed, each foregoing a ten-second look and saving ten minutes on the freeway.

Let $W(\lambda,\mu)$ denote the expected waiting time and let $W_q(\lambda,\mu)$ denote the expected queueing time, given the rates λ and μ.[6] A customer's choice of service rate does not affect his queueing time. Therefore, his decision amounts to selecting a service rate μ_e which minimizes $\frac{C}{\mu}+h(\mu)$ over $\mu > 0$, and μ_e satisfies

$$h'(\mu_e) = \frac{C}{\mu_e^2}\,.$$

Note that μ_e is not a function of λ. In contrast, the social objective (given that the arrival rate is λ) is to set μ to minimize $CW(\lambda,\mu)+h(\mu)$. Given the assumption on the function $h(\mu)$, the optimal service rate, $\mu^*(\lambda)$, is uniquely determined by the first-order condition

$$h'(\mu^*(\lambda)) = -C\frac{\partial W(\lambda,\mu^*(\lambda))}{\partial\mu}. \tag{8.8}$$

By the convexity of $W(\lambda,\mu)$ in μ, and the assumptions on the cost function h:

- $\mu^*(\lambda)$ increases in λ;
- $\mu^*(\lambda) \geq \mu_e$ for $\lambda \geq 0$.

To induce the socially optimal behavior in equilibrium, Ha suggested a price scheme which consists of both a *fixed admission fee* α and a *variable service fee* β proportional to the realized time of service. (If it is possible to observe the service rate chosen by a customer, the variable part of the price can be made proportional to the expected service time.) The goal of these fees is to attain equilibrium with the optimal arrival and service rates.

We first determine the variable cost, β. The customer's choice of μ minimizes $h(\mu)+\frac{C+\beta}{\mu}$. This is a strictly convex function of μ and the unique minimizer satisfies

$$h'(\mu) = \frac{C+\beta}{\mu^2}. \tag{8.9}$$

Since the expected number of customers in the system, $L(\lambda,\mu)$, is a function of λ and μ only through $\frac{\lambda}{\mu}$, $\lambda\frac{\partial L}{\partial\lambda}+\mu\frac{\partial L}{\partial\mu}=0$. With $L=\lambda W$ this becomes

$$W+\lambda\frac{\partial W}{\partial\lambda}+\mu\frac{\partial W}{\partial\mu}=0. \tag{8.10}$$

[6] A more formal approach would require to define the expected waiting time for any strategy profile selected by the customers but these definitions are sufficient since we only consider symmetric profiles.

Let λ^* and μ^* be the arrival and service rates which jointly maximize social welfare. By substituting (8.9) and (8.10) into (8.8) we obtain the following theorem:

THEOREM 8.2 *Suppose that $\lambda = \lambda^*$ and that the service charge per unit of actual service is*

$$\beta = \mu^* C \left(W_q(\lambda^*, \mu^*) + \lambda^* \frac{\partial W(\lambda^*, \mu^*)}{\partial \lambda} \right).$$

Then the resulting equilibrium service rate equals μ^*.

We now determine the optimal admission fee, α. The social objective is to maximize $V(\lambda) - C\lambda W(\lambda, \mu) - \lambda h(\mu)$. This function is strictly concave in λ and hence its unique maximizer is determined by the first-order condition

$$V'(\lambda) = CW(\lambda, \mu) + C\lambda \frac{\partial W(\lambda, \mu)}{\partial \lambda} + h(\mu).$$

In equilibrium, the arrival rate is such that $V'(\lambda)$ equals the expected full price:

$$\alpha + \frac{\beta}{\mu} + CW(\lambda, \mu) + h(\mu). \tag{8.11}$$

Thus we obtain the following theorem:

THEOREM 8.3 *Suppose that an admission fee of $\alpha = -CW_q(\lambda^*, \mu^*)$ and a variable fee of β as stated in Theorem 8.2 are imposed. Then the socially optimal solution (λ^*, μ^*) defines an equilibrium.*

6.2. Multi-class extensions

Ha [63] extended the model of [62], assuming that the demand process consists of m classes that differ by their aggregate utility functions $V_i(\lambda_i)$, cost functions $h_i(\mu_i)$, and time values C_i, $i = 1, \ldots, m$. We will assume that class identities are unobservable.[7] It is assumed that for each class i, the values of $V_i'(0)$ and $h_i(0)$ are sufficiently large to guarantee interior solutions. Let $\underline{\lambda} = (\lambda_1, \ldots, \lambda_m)$ and $\underline{\mu} = (\mu_1, \ldots, \mu_m)$. The social objective is to maximize

$$\sum_{i=1}^{m} \left[V_i(\lambda_i) - \lambda_i C_i W_i(\underline{\lambda}, \underline{\mu}) - \lambda_i h_i(\mu_i) \right] \tag{8.12}$$

[7] Ha also considers the case when class identities are observable.

with respect to $\underline{\lambda}$ and $\underline{\mu}$, where $W_i(\underline{\lambda}, \underline{\mu})$ is the expected waiting time of an i-customer given the rates $\underline{\lambda}$ and $\underline{\mu}$.

Let $\underline{\lambda}^*$ and $\underline{\mu}^*$ be the vectors of arrival and service rates which jointly maximize (8.12). The first-order conditions are for $i = 1, \ldots, m$,

$$\sum_{j=1}^{m} C_j \lambda_j^* \frac{\partial W_j(\underline{\lambda}^*, \underline{\mu}^*)}{\partial \mu_i} + \lambda_i^* h_i'(\mu_i^*) = 0, \tag{8.13}$$

and

$$V_i'(\lambda_i^*) = C_i W_i(\underline{\lambda}^*, \underline{\mu}^*) + \sum_{j=1}^{m} C_j \lambda_j^* \frac{\partial W_j(\underline{\lambda}^*, \underline{\mu}^*)}{\partial \lambda_i} + h_i(\mu_i^*). \tag{8.14}$$

Let t be an observable measure associated with a customer's service rate. Specifically, Ha considered two service disciplines, FCFS in which t is the time in service, and EPS in which t is the time in the system. Let τ be a random variable denoting the realization of t. Suppose that the server charges customers a price $p(t)$. Let $W_i(\underline{\lambda}, \underline{\mu}, \mu)$ and $E[p(\tau_i)|\underline{\lambda}, \underline{\mu}, \mu]$ denote the expected waiting time and price, respectively, incurred by an i-customer who chose a service rate μ. The objective of an i-customer is to choose a μ value that minimizes

$$E[p(\tau_i)|\underline{\lambda}, \underline{\mu}, \mu] + C_i W_i(\underline{\lambda}, \underline{\mu}, \mu) + h_i(\mu).$$

The first-order conditions are that for $i = 1, \ldots, m$,

$$C_i \frac{\partial W_i(\underline{\lambda}, \underline{\mu}, \mu)}{\partial \mu} + \frac{\partial E(p(\tau_i)|\underline{\lambda}, \underline{\mu}, \mu)}{\partial \mu} + h_i'(\mu) = 0. \tag{8.15}$$

In a symmetric equilibrium, (8.15) holds with $\mu = \mu_i$ for $i = 1, \ldots, m$. The arrival rates λ_i are determined by

$$V_i'(\lambda_i) = E[p(\tau_i)|\underline{\lambda}, \underline{\mu}, \mu_i] + C_i W_i(\underline{\lambda}, \underline{\mu}) + h(\mu_i). \tag{8.16}$$

Comparing (8.13), (8.14), (8.15), and (8.16), the price function $p(t)$ will induce optimal arrival and service rates in equilibrium if for $i = 1, \ldots, m$,

$$\frac{\partial E[p(\tau_i)|\underline{\lambda}^*, \underline{\mu}^*, \mu_i^*]}{\partial \mu_i} = \frac{1}{\lambda_i^*} \sum_{j=1}^{m} C_j \lambda_j^* \frac{\partial W_j(\underline{\lambda}^*, \underline{\mu}^*)}{\partial \mu_i} - C_i \frac{\partial W_i(\underline{\lambda}^*, \underline{\mu}^*, \mu_i^*)}{\partial \mu_i} \equiv E_i^s, \tag{8.17}$$

and

$$E[p(\tau_i)|\underline{\lambda}^*, \underline{\mu}^*] = \sum_{j=1}^{m} C_j \lambda_j^* \frac{\partial W_j(\underline{\lambda}^*, \underline{\mu}^*)}{\partial \lambda_i} \equiv E_i^a. \tag{8.18}$$

Ha interpreted E_i^s and E_i^a as service and admission externalities, respectively.

Ha applied (8.17) and (8.18) to the following two models:

- Consider an $M/G/s$ system with egalitarian processor sharing (EPS): a server is dedicated to a customer if the number of customers does not exceed s; otherwise, the service capacity is allocated equally among the customers in the system.

  Important features of the EPS model are:

  - The service externalities are identical across all classes, that is, $E_i^s = E^s$ for $i = 1, \ldots, m$.
  - The admission externalities equal the service externalities, that is, $E_i^a = \frac{E^s}{\mu_i^*}$, for $i = 1, \ldots, m$.

  Consequently, a single undifferentiated price per unit of time in the system can be applied to optimally regulate customers' behavior in this model of heterogeneous customer classes. For some constant β, $p(t) = \beta t$ induces an equilibrium in which customers in each class make the systemwide optimal admission decisions and those who join are induced to select the optimal service requirement intended for their class.

- Consider an $M/G/1$ FCFS system. In this case, Ha proved that there are constants β and γ such that the pricing function $p(t) = \beta t + \gamma t^2$ induces the optimal behavior in equilibrium.

7. Competition among servers

Chen and Wan [35] considered a model of competition between two identical servers who maximize their profits by choosing both prices and service capacities. They assume that customers are homogeneous and show that four cases are possible:

- A unique equilibrium in which none of the firms operates in the market.
- Countable equilibria in which one of the firms captures the whole market.
- Countable equilibria in which each firm captures half of the market.
- A continuum set of equilibria in which the firms divide the market into different shares.

Cachon and Harker [31] considered an unobservable queueing system with two servers who differ by some parameter. Customers have preferences over the value of this parameter and incur a cost proportional to the difference between their desired value and that offered by the firm they selected. This is a variation of the classic Hotelling model [82]. It is assumed that servers compete by offering prices and waiting cost guarantees (by controlling the service rate). When the servers have identical cost parameters there is a unique equilibrium in which, in contrast to the model of Levhari and Luski [113, 104] (see, §7.1), if the firm's costs are identical then the firms adopt identical strategies.

Kalai, Kamien and Rubinovitch [85] considered a model in which servers choose their service rate in a competitive environment, in order to capture a larger market share. The model assumes a single queue. A customer who encounters free servers randomly chooses one, independently of their service rates. Otherwise, he joins the queue. When service ends and the queue is not empty, the first customer in the queue moves to obtain service. This model makes the problematic assumption that customers do not give priority to faster servers. One could argue that this is a result of ignorance, but clearly the faster servers have an incentive to advertise the distinction between themselves and the slower servers.

Gilbert and Weng [57] built on Kalai, Kamien and Rubinovitch [85], assuming that two servers are controlled by a coordinating agency. Each server selects a profit maximizing service rate, but the agency allocates customers to servers and compensates servers accordingly in order to achieve given expected waiting times at a minimum cost. The analysis shows that the coordinating agency may prefer a separate queue allocation scheme to a common queue.

8. Capacity expansion

Capacity expansion is an important subject of operations research (see the survey by Luss [114]), but not much has been done on related queueing models. We describe below examples of such models.

Consider an unobservable $M/M/s$ system, and assume that the population of potential customers grows with time, so that at time t the potential arrival rate is $\Lambda(t)$ where $\Lambda(t)$ is monotone increasing. The facility manager has to decide on the admission fee and on a sequence of instants $t_1, t_2, \ldots$ when it is worth adding a new server (which is associated with either increased operation costs or capacity expansion costs). The objective of the manager is to maximize the discounted present value of the system's profits or social welfare.

Alternatively, consider a class model of heterogeneous customers. The arrival rate of i-customers at time t is determined by the equilibrium condition $V_i'(\lambda_i) = P_i$ where P_i denote their full price. Suppose that the population of potential customers grows uniformly with time, so that at time t the marginal value function is $\Lambda(t)V_i'$ where $\Lambda(t)$ is monotone increasing. The facility manager has to decide on both the admission fees imposed on each customer class and on a sequence of instants $t_1, t_2, \ldots$ when it is worth adding a new server.

A variation of this model was considered by Oum and Zhang [137]. In their model, the admission fee imposed on an i-customer is equal to the externalities such a customer imposes on the system. They conducted numerical experiments with specific data and growth function $\Lambda(t) = \Lambda[1 + (\alpha t)^\beta]$. They also applied a strong assumption that the total arrival rate is a function of the average (over classes) expected full price.

9. Related literature

- Balachandran and Radhakrishnan [19] considered an $M/M/1$ model with fixed arrival rates $\lambda_1, \ldots, \lambda_m$ of customer classes. The classes jointly decide on μ and on an allocation of $c(\mu)$ among them, so that class i pays a fraction γ_i of $c(\mu)$. The waiting cost parameter for i-customers is C_i. Each class is considered an entity of its own, bearing all the waiting costs of its customers and its fraction of $c(\mu)$.

  The socially optimal service rate μ satisfies the first-order condition

$$\frac{\partial}{\partial \mu}\left(\sum_i C_i \frac{\lambda_i}{\mu - \sum_j \lambda_j} + c(\mu)\right) = -\frac{\sum_i C_i \lambda_i}{(\mu - \sum_j \lambda_j)^2} + \frac{\partial c(\mu)}{\partial \mu} = 0.$$

  Under individual class optimization, the service rate is determined by

$$-\frac{C_i \lambda_i}{(\mu - \sum_j \lambda_j)^2} + \gamma_i \frac{\partial c(\mu)}{\partial \mu} = 0, \quad 1 \le i \le m.$$

  Consider an allocation with fractions

$$\gamma_i = \frac{C_i \lambda_i}{\sum_j C_j \lambda_j}, \quad 1 \le i \le m.$$

  With these fractions, the socially optimal service rate is optimal for every class.

- Chen and Frank [33] considered a long-run model of a profit maximizing server in which the cost of maintaining a service rate μ is

$b\mu$ per unit of time, and the cost of serving a customer is r. They observed that in this model, if a positive profit is possible, then the server will select a processing rate μ and an admission fee p such that all the potential arrivals will be served. Since the maximum fee that can be charged while maintaining the arrival rate Λ is $p = R - \frac{C}{\mu-\Lambda}$, the problem becomes

$$\max_{r<p<R-C/\mu} (p-r)\Lambda - b\Big(\Lambda + \frac{C}{R-p}\Big).$$

The solution is $p^* = R - \sqrt{\frac{Cb}{\Lambda}}$ and $\mu^* = \Lambda + \sqrt{\frac{C\Lambda}{b}}$. Chen and Frank observed that:

- The solution does not vary with the cost, r, of serving a customer. This cost only determines whether a positive profit is possible. The condition for a positive profit is $r < R - \sqrt{\frac{Cb}{\Lambda}}$. Note that the left-hand side is the optimal admission fee, p^*.
- The firm responds to an increase in Λ by increasing μ and p.
- As in the short-run model (§3.1), the profit-maximizing solution is socially optimal.

■ Ittig [83] assumed a revenue function that depends on the arrival rate and a cost function that depends on the number of servers in an $M/M/s$ queue or on the service rate in an $M/M/1$ queue. For several demand functions, Ittig computed the optimal number of servers and the service rate which maximize social welfare.

■ The model by So and Song [158] assumes that the demand for service is a function of two parameters, the price and the α-percentile of the waiting time distribution (the "delivery time guarantee"), for some predetermined α. In particular, the demand function is $\lambda p^{-\alpha} x^{-\beta}$, where p is the price of service, and the probability that the waiting time is at most x equals α. The server has the option of increasing the service rate at a linear cost. The paper characterizes the optimal price and capacity selection for an $M/M/1$ system.

■ In §3.4 we described an interesting type of equilibrium inefficiency discovered by Balachandran and Schaefer [20]: when the potential population of customers consists of classes that differ by their cost/reward ratios, a single class arrives ("dominates") in equilibrium and it is not necessarily the socially desired class. Balachandran and Radhakrishnan [19] proved that this inefficiency does not exist in the long-run model when the cost of operating the server is linear in the service rate.

- Balachandran and Srinidhi [25] considered an $M/G/1$ model in which the cost of operating a server at rate μ given an arrival rate λ is proportional to $e^{-\frac{\lambda}{\mu}}$.[8] They show that for this cost function, the short- and long-run first-order optimality conditions for social optimality coincide. However, in the short-run model they ignored the dependence of the operating costs on the arrival rate.

[8]They provide no justification for using this function.

References

[1] Ackere, A. van and P. Ninios (1993) "Simulation and queueing theory applied to a single-server queue with advertising and balking," *Journal of the Operational Research Society* 44, 407-414.

[2] Adiri, I. and U. Yechiali (1974) "Optimal priority purchasing and pricing decisions in nonmonpoly and monopoly queues," *Operations Research* 22, 1051-1066.

[3] Afèche, P. and H. Mendelson (2001) "Priority auctions vs. uniform pricing in queueing systems with a generalized delay cost structure."

[4] Agastya, M. (2001), personal communication.

[5] Agrawala, A.K., E. G. Coffman, Jr., M. R. Garey and S. K. Tripathi (1984) "A stochastic optimization algorithm minimizing expected flow times on uniform processors," *IEEE Transactions on Computers* C-33, 351-356.

[6] Alexeev, M. (1989) "A note on privileges in a queue-rationed CPE with black markets," *Journal of Economic Theory* 47, 422-430.

[7] Alperstein, H. (1988) "Optimal pricing for the service facility offering a set of priority prices," *Management Science* 34, 666-671.

[8] E. Altman, T. Boulogne, R. El Azouzi and T. Jimenez (2000) "A survey on networking games in telecommunication."

[9] Altman, E. and R. Hassin (2001), "Non-threshold equilibrium for customers joining an $M/G/1$ queue."

[10] Altman, E. and N. Shimkin (1998) "Individual equilibrium and learning in processor sharing systems," *Operations Research* 46, 776-784.

[11] Armony, M. and M. Haviv (2002) "Price and delay competition in make-to-order operations," *European Journal of Operational Research.*

[12] Arnott, R., A. de Palma, and R. Lindsey (1999) "Information and time-of-usage decisions in the bottleneck model with stochastic capacity and demand," *European Economic Review* 43, 525-548.

[13] Assaf, D. and M. Haviv (1990) "Reneging from time sharing and random queues," *Mathematics of Operations Research* 15, 129-138.

[14] Atkinson, J.B. (1996) "A note on a queueing optimization problem," *Journal of the Operational Research Society* 47, 463-467.

[15] Balachandran, K.R. (1972) "Purchasing priorities in queues," *Management Science* 18, 319-326.

[16] Balachandran, K.R. (1991) "Incentive and regulation in queues," *Lecture Notes in Economics and Mathematical Systems*, M.J. Beckmann, M.N. Gopalan, and R. Subramanian (Eds.), 370, 162-176.

[17] Balachandran, K.R. and J.C. Lukens (1976) "Stable pricing in service systems," *Zeitschrift für Operations Research* 20, 189-201.

[18] Balachandran, K.R. and S. Radhakrishnan (1994) "Extension to class dominance characteristics," *Management Science*, 40, 1353-1360.

[19] Balachandran, K.R. and S. Radhakrishnan (1996) "Cost of congestion, operational efficiency and management accounting," *European Journal of Operational Research* 89, 237-245.

[20] Balachandran, K.R. and M.E. Schaefer (1979) "Class dominance characteristics at a service facility," *Econometrica* 47, 515-519.

[21] Balachandran, K.R. and M.E. Schaefer (1979) "Regulation by price of arrivals to a congested facility," *Cahiers du C.E.R.O.* 21, 149-158.

[22] Balachandran, K.R. and M.E. Schaefer (1980) "Public and private optimization at a service facility with approximate information on congestion," *European Journal of Operational Research* 4, 195-202.

[23] Balachandran, K.R. and B. Srinidhi (1987) "A rationale for fixed charge application," *Journal of Accounting, Auditing and Finance* 14, 151-169.

[24] Balachandran, K.R. and B. Srinidhi (1988) "A stable cost application scheme for service center usage," *Journal of Accounting, Auditing and Finance* 15, 87-99.

[25] Balachandran, K.R. and B. Srinidhi (1990) "A note on cost allocation, opportunity cost and optimal utilization," *Journal of Business, Finance and Accounting*, 579-584.

[26] Barzel Y. (1974) "A theory of rationing by waiting," *Journal of Law and Economics* 17, 73-95.

[27] Bell, C.E. and S. Stidham, Jr. (1983) "Individual versus social optimization in allocation of customers to alternative servers," *Management Science* 29, 831-839.

[28] Ben-Shahar, I., A. Orda, and N. Shimkin (2000) "Dynamic service sharing with heterogeneous preferences," *Queueing Systems: Theory and Applications*, 35, 83-103.

[29] Bradford, R.M. (1996) "Incentive compatible pricing and routing policies for multi-server queues," *European Journal of Operational Research*, 89, 226-236.

[30] Braess, D. (1968) "Über ein paradoxon aus der verkehrsplanung," *Unternehmensforschung*, 12, 258-268.

[31] Cachon, G.P and P.T. Harker (1999) "Service competition, outsourcing and co-production in a queueing game."

[32] Calvert, B., W. Solomon, and I. Ziedins (1997) "Braess's paradox in a queueing network with state dependent routing," *Journal of Applied Probability*, 34, 134-154.

[33] Chen, H. and M. Frank, (1994) "Monopoly pricing when customers queue," Working Paper, Faculty of Commerce and Business Administration, University of British Columbia.

[34] Chen, H. and M. Frank, (2001) "State dependent pricing with a queue," *IIE Transactions* 33, 847-860.

[35] Chen, H. and Y-w. Wan, (2002) "Price competition of make-to-order firms."

[36] Cohen, J.E. and F.P. Kelly (1990) "A paradox of congestion in a queueing network," *Journal of Applied Probability* 27, 730-734.

[37] Cox, D. and Smith, W. (1961) *Queues*, Methuen and Company, London.

[38] Crawford, V.P. (1995) "Adaptive dynamics in coordination games," *Econometrica* 63, 103-143.

[39] Daniel, J. (1995) "Congestion pricing and capacity of large hub airports: a bottleneck model with stochastic queues," *Econometrica* 63, 327-370.

[40] Davidson, C. (1988) "Equilibrium in servicing industries: an economic application of queueing theory," *Journal of Business* 61, 347-367.

[41] De Vany, A. (1976) "Uncertainty, waiting time, and capacity utilization: a stochastic theory of product quality," *Journal of Political Economy* 84, 523-541.

[42] De Vany, A.S. and T.R. Saving (1983) "The economics of quality," *Journal of Political Economy* 91, 979-1000.

[43] Deacon, R.T. and J. Sonstelie (1985) "Rationing by waiting and the value of time: results from a natural experiment," *Journal of Political Economy* 93, 637-647.

[44] Dewan S. and H. Mendelson (1990) "User delay costs and internal pricing for a service facility," *Management Science* 36, 1502-1517.

[45] Dolan R.J. (1978) "Incentive mechanisms for priority queueing problems," *Bell Journal of Economics* 9, 421-436.

[46] Donaldson, D. and B.C. Eaton (1981) "Patience, more than its own reward: a note on price discrimination," *Canadian Journal of Economics* 14, 93-105.

[47] Edelson, N.M. and K. Hildebrand (1975) "Congestion tolls for Poisson queueing processes," *Econometrica* 43, 81-92.

[48] Elcan, A. (1994) "Optimal customer return rate for an $M/M/1$ queueing system with retrials," *Probability in the Engineering and Informational Sciences* 8, 521-539.

[49] Falin, G. (1990) "A survey of retrial queues," *Queueing Systems: Theory and Applications* 7, 127-168.

[50] Fayolle, G. and R. Iasnogorodski (1979) "Two couple processors: the reduction to a Riemann-Hilbert problem," *Z. Wahrscheinlichkeits Theorie*, 47, 325-351.

[51] Fayolle, G., I. Mitrani and R. Iasnogorodski (1980), "Sharing a processor among many job classes," *Journal of the Association for Computing Machinery* 27, 519-532.

[52] Federgruen, A. and H. Groenevelt (1988) "Characterization and optimization of achievable performance in general queueing systems," *Operations Research* 36, 733-741.

[53] Feller, W. (1968), *Introduction to Probability and Its Applications, Vol. 1, 3rd edition*, John Wiley & Sons, New York.

[54] Friedman, E.J. and A.S. Landsberg (1993) "Short-run dynamics of multi-class queues," *Operations Research Letters* 14, 221-229.

[55] Ghanem, S.B. (1975) "Computing central optimization by a pricing priority policy," *IBM Systems Journal* 14, 272-292.

[56] Gibbens, R.J. and F.P. Kelly (1999) "Resource pricing and the evolution of congestion control," *Automatica* 35, 1969-1985.

[57] Gilbert, S.M. and Z.K. Weng (1998) "Incentive effects favor nonconsolidating queues in a service system: the principal agent perspective," *Management Science* 44, 1662-1669.

[58] Glazer, A. and R, Hassin (1983) "Search among queues," unpublished report.

[59] Glazer, A. and R. Hassin (1983) "$?/M/1$: On the equilibrium distribution of customer arrivals," *European Journal of Operational Research* 13, 146-150.

[60] Glazer, A. and R. Hassin (1986) "Stable priority purchasing in queues," *Operations Research Letters* 4, 285-288.

[61] Glazer, A. and R. Hassin (1987) "Equilibrium arrivals in queues with balk service at scheduled times," *Transportation Science* 21, 273-278.

[62] A.Y. Ha (1998) "Incentive-compatible pricing for a service facility with joint production and congestion externalities," *Management Science* 44, 1623-1636.

[63] A.Y. Ha (2001) "Optimal pricing that coordinate queues with customer-chosen service requirements," *Management Science* 47, 915-930.

[64] G. Hardin (1968) "The tragedy of the commons," *Science* 162, 1243-48.

[65] Hassin, R. (1985) "On the optimality of first come last served queues," *Econometrica* 53, 201-202.

[66] Hassin, R. (1986) "Consumer information in markets with random products quality: The case of queues and balking," *Econometrica* 54, 1185-1195.

[67] Hassin, R. (1995) "Decentralized regulation of a queue," *Management Science* 41, 163-173.

[68] Hassin, R. (1996) "On the advantage of being the first server," *Management Science* 42, 618-623.

[69] Hassin, R. and M. Haviv (1994) "Equilibrium strategies and the value of information in a two line queueing system with threshold jockeying," *Communications in Statistics - Stochastic Models* 10, 415-436.

[70] Hassin, R. and M. Haviv (1995) "Equilibrium strategies for queues with impatient customers," *Operations Research Letters* 17, 41-45.

[71] Hassin, R. and M. Haviv (1996) "Optimal and equilibrium retrial rates in a busy system," *Probability in the Engineering and Informational Sciences* 10, 223-227.

[72] Hassin, R. and M. Haviv (1997) "Equilibrium threshold strategies: the case of queues with priorities," *Operations Research,* 45, 966-973.

[73] R. Hassin and Haviv, M. (2002) "Nash equilibrium and subgame perfection: the case of observable queues," *Annals of Operations Research.*

[74] Hassin, R. and M. Henig (1986) "Control of arrivals and departures in a state-dependent input-output system," *Operations Research Letters* 5, 33-36.

[75] Haviv, H. (1991) "Stable strategies for processor sharing systems," *European Journal of Operational Research* 52, 103-106.

[76] Haviv, M. (2001) "The Aumann-Shaply price mechanism for allocating costs in congestion systems," *Operations Research Letters,* 29, 211-215.

[77] Haviv, M. and M.L. Puterman (1998) "Bias optimality in controlled queueing systems," *Journal of Applied Probability* 35, 136-150.

[78] Haviv, M. and Y. Ritov (2001) "Homogeneous customers renege from invisible queues at random times under deteriorating waiting conditions," *Queueing Systems: Theory and Applications,* 38, 495-508.

[79] Haviv, M. and J. van der Wal (1997) "Equilibrium strategies for processor sharing and queues with relative priorities," *Probability in the Engineering and Informational Sciences* 11, 403-412.

[80] Hlynka, M., D.A. Stanford, W.H. Poon, and T. Wang (1994) "Observing queues before joining," *Operations Research* 42, 365-371.

[81] Holt, C.A. Jr. and R. Sherman (1982) "Waiting-line auctions," *Journal of Political Economy* 90, 280-294.

[82] Hotelling, H. (1929) "Stability in competition," *Economic Journal* 39, 41-57.

[83] Ittig, P.T. (1994) "Planning service capacity when demand is sensitive to delay," *Decision Sciences* 25, 541-559.

[84] Johansen, S.G. and S. Stidham, Jr. (1980) "Control of arrivals to a stochastic input-output system," *Advances in Applied Probability* 12, 972-999.

[85] Kalai, E., M. I. Kamien and M. Rubinovitch, (1992) "Optimal service speeds in a competitive environment," *Management Science* 38, 1154-1163.

[86] Kingman, J.F.C, (1961) "Two similar queues in parallel," *Annals of Mathematical Statistics* 32, 1314-1323.

[87] Kingman, J.F.C. (1961) "The effect of queue discipline on waiting time variance," *Proceeding of the Cambridge Philosophical Society* 63, 163-164.

[88] Kleinrock, L. (1967) "Optimal bribing for queue position," *Operations Research* 15, 304-318.

[89] Kleinrock, L. (1976) *Queueing Systems, Vol. 2: Computer Applications*, John Wiley & Sons, New York.

[90] Knudsen, N. C. (1972) "Individual and social optimization in a multi-server queue with a general cost-benefit structure," *Econometrica* 40, 515-528.

[91] Koenigsberg, E. (1980) "Uncertainty, capacity and market share in oligopoly: a stochastic theory of product quality," *Journal of Business* 53, 151-164.

[92] Koenigsberg, E. (1985) "Queue systems with balking: a stochastic model of price discrimination," *R.A.I.R.O. Recherche Opérationnelle* 19, 209-219.

[93] Kosten, L. (1973) *Stochastic Theory of Service Systems*, Pergamon Press, Oxford.

[94] Krueger, A (1974) "The political economy of the rent-seeking society," *American Economic Review* 64, 291-303.

[95] Kulkarni, V.G. (1983) "On queueing systems with retrials," *Journal of Applied Probability* 20, 380-389.

[96] Kulkarni V.G. (1983) "A game theoretic model for two types of customers competing for service," *Operations Research Letters* 2, 119-122.

[97] Kumar, P.R. and J. Walrand (1985) "Individually optimal routing in parallel systems," *Journal of Applied Probability* 22, 989-995.

[98] Landsburg, S.E. (2001) "The first one now will later be last," *Slate Archives*, http://slate.msn.com/economics/01-02-08/economics.asp.

[99] Larsen, C. (1998) "Investigating sensitivity and the impact of information on pricing decisions in an $M/M/1/\infty$ queueing model," *International Journal of Production Economics* 56-57, 365-377.

[100] Larson, R.C. (1987) "Perspectives on queues: social justice and the psychology of queueing," *Operations Research* 35, (1987) 895-909.

[101] Lederer P.J. and L. Li (1997) "Pricing, production, scheduling, and delivery time competition," *Operations Research* 45, 407-420.

[102] Lee, H.L. and M.A. Cohen (1985) "Multi-agent customer allocation in a stochastic service system," *Management Science* 31, 752-763.

[103] Leeman, W.A. (1964) "The reduction of queues through the use of price," *Operations Research* 12, 783-785.

[104] Levhari, D. and I. Luski (1978) "Duopoly pricing and waiting lines," *European Economic Review* 11, 17-35.

[105] Levhari, D. and E. Sheshinski (1974) "The economics of queues: a brief survey," 195-212 in *Essays in Economic Behavior Under Uncertainty* M. Balch, D. McFadden, and S. Wo (Eds.) Elsevier, New York.

[106] Levy, A. and H. Levy (1991) "Lock and no-lock mortgage plans: is it only a matter of risk shifting?" *Operations Research Letters* 10, 233-240.

[107] Li, L. and Y.S. Lee (1994) "Pricing and delivery-time performance in a competitive environment," *Management Science* 40, 633-646.

[108] Lin, K.Y. and S.M. Ross (2001) "Admission control with incomplete information of a queueing system."

[109] Lippman, S.A. and S. Stidham, Jr. (1977) "Individual versus social optimization in exponential congested systems," *Operations Research* 25, 233-247.

[110] Littlechild, S.C (1974) "Optimal arrival rate in a simple queueing system," *International Journal of Production Research* 12, 391-397.

[111] Loch, C. (1991) "Pricing in markets sensitive to delay," Ph.D. Dissertation, Stanford University.

[112] Lui, F.T (1985) "An equilibrium queueing model of bribery," *Journal of Political Economy* 93, 760-781.

[113] Luski, I. (1976) "On partial equilibrium in a queueing system with two servers," *The Review of Economic Studies* 43, 519-525.

[114] Luss, H. (1982) "Operations research and capacity expansion problems: a survey," *Operations Research* 30, 907-947.

[115] MacKie-Mason, J. and H.R. Varian (1995) "Pricing the Internet," in *Public Access to the Internet,* (Brian Kahin and James Keller, eds.), The MIT Press, Cambridge, MA. 269-314.

[116] MacKie-Mason, J.K. and H.R Varian (1995) "Pricing congestible network resources," *IEEE Journal on Selected Areas in Communication* 13, 1141-1149.

[117] Mandelbaum, A. and N. Shimkin (2000) "A model for rational abandonments from invisible queues," *Queueing Systems: Theory and Applications* 6, 141-173.

[118] Mandelbaum, A. and U. Yechiali (1983) "Optimal entering rules for a customer with wait option at an $M/G/1$ queue," *Management Science* 29, 174-187.

[119] Marchand, M.G. (1974) "Priority pricing," *Management Science* 26, 1131-1140.

[120] Masuda, Y. and S. Whang (1999) "Dynamic pricing for network service: equilibrium and stability," *Management Science* 45, 857-869.

[121] Masuda, Y. and S. Whang (2000) "Capacity management in decentralized networks."

[122] Maynard-Smith, J. (1982) *Evolution and Game Theory*, Cambridge University Press, Cambridge.

[123] Mendelson, H. (1985) "Pricing services: queueing effects," *Communications of the ACM* 28, 312-321.

[124] Mendelson, H. and S. Whang (1990) "Optimal incentive-compatible priority pricing for the $M/M/1$ queue," *Operations Research* 38, 870-883.

[125] Mendelson, H. and Y. Yechiali (1981) "Controlling the $GI/M/1$ queue by conditional acceptance of customers," *European Journal of Operational Research* 7, 77-85.

[126] Mieghem, J.A. van (1995) "Dynamic scheduling with convex delay costs: the generalized $c\mu$ rule," *Annals of Applied Probability* 5, 809-833.

[127] Mieghem, J.A. van (2000) "Price and service discrimination in queueing systems: incentive compatibility of $Gc\mu$ scheduling," *Management Science* 46, 1249-1267.

[128] Miller, B.L, and A.G. Buckman (1987) "Cost allocation and opportunity costs," *Management Science* 33, 626-639.

[129] Mills, D. E. (1981) "Ownership arrangements and congestion-prone facilities," *The American Economic Review* 71, 493-502.

[130] Morrison, S.A. (1986) "A Survey of Road Pricing," *Transportation Research* 20A, 87-97.

[131] Myrdal, G. (1968) *Asian Drama: An Inquiry into the Poverty of Nations*, Pantheon, New York.

[132] Nalebuff, B. (1989) "The arbitrage mirage, waitwatchers, and more," *Journal of Economic Perspectives* 3, 1656-174.

[133] Naor, P. (1969) "The regulation of queue size by levying tolls," *Econometrica* 37, 15-24.

[134] Neuts, M.F. (1981), *Matrix-Geometric Solutions in Stochastic Models*, The Johns Hopkins University Press, Baltimore.

[135] Nichols, D., E. Smolensky, and T.N. Tideman (1971) "Discriminating by waiting time in merit goods," *American Economic Review* 61, 312-323.

[136] Olson, M.A. (1992) "Queues when balking is strategic," ICES working paper #1, George Mason University, Fairfax VA, http://mason.gmu.edu/ molson2.

[137] Oum, T.H. and Y. Zhang (1990) "Airport Pricing: congestion tolls, Lumpy investment, and cost recovery," *Journal of Public Economics* 43, 353-374.

[138] Parra-Frutos, I. and J.Aranda-Gallego (1999) "Multiproduct monopoly: a queueing approach," *Applied Economics* 31, 565-576.

[139] Page, E (1972) *Queueing Theory in OR*, Crane Russak and Company, Inc. New York.

[140] Png, I.P.L. and D. Reitman (1994) "Service time competition," *RAND Journal of Economics* 25, 619-634.

[141] Polterovich, V. (1993) "Rationing, queues, and black markets," *Econometrica* 61, 1-28.

[142] Radhakrishnan, S. and K.R. Balachandran (1995) "Delay cost and incentive schemes for multiple users," *Management Science* 41, 646-652.

[143] Rao, S. and E.R. Petersen (1998) "Optimal pricing of priority services," *Operations Research* 46, 46-56.

[144] Rapoport, A., W.E. Stein, J.E. Parco, and D.A. Seale (2001) "Strategic play in single-server queues with endogenously determined arrival times."

[145] Rashid, S. (1981) "Public utilities in egalitarian LDC's: the role of bribery in achieving Pareto efficiency," *Kyklos* 34, 448-460.

[146] Reitman, D. (1991) "Endogenous quality differentiation in congested markets," *The Journal of Industrial Economics* 39, 621-647.

[147] Rosenblum, D.M. (1992) "Allocation of waiting time by trading in position on a $G/M/s$ queue," *Operations Research* 40, S338-S342.

[148] Ross, S.M. (1983), *Stochastic Processes*, John Wiley & Sons, New York.

[149] Rue, R. C. and M. Rosenshine (1981) "Some properties of optimal control policies for entry to an $M/M/1$ queue," *Naval Research Logistics Quarterly* 28, 525-532.

[150] Rump, C.M. and S. Stidham Jr. (1998) "Stability and chaos in input pricing for a service facility with adaptive customer response to congestion," *Management Science* 44, 246-261.

[151] Samuelson, W.F. (1985) "Competitive bidding with entry costs," *Economics Letters* 17, 53-57.

[152] Sanders B.A. (1985) "A private good/public good decomposition for optimal flow control of an $M/M/1$ queue," *IEEE Transactions on Automatic Control* AC-30, 1143-1145.

[153] Schelling T.C. (1978) *Micromotives and Macrobehavior*, Norton & Company, New York.

[154] Schroeter, J. R. (1982) "The costs of concealing the customer queue," working paper EC-118, Bureau of Business and Economic Research, Arizona State University.

[155] Shimkin, N. and A. Mandelbaum (2002) "Rational abandonment from tele-queues: nonlinear waiting costs with heterogeneous preferences."

[156] Shy, O. (2001) *The Economics of Networks Industries*, Cambridge University Press, Cambridge, UK.

[157] Simonovits, A. (1976) "Self- and social optimization in queues," *Studia Scientiarum Mathematicarum Hungarica* 11, 131-138.

[158] So, K.C. and J-S Song (1998) "Price, delivery time guarantees and capacity selection," *European Journal of Operational Research* 111, 28-49.

[159] Spence, A.M. (1975) "Monopoly, quality, and regulation," *The Bell Journal of Economics* 6, 417-429.

[160] Stahl, D.O. and M. Alexeev (1985) "The influence of black markets on a queue-rationed centrally planned economy," *Journal of Economic Theory* 35, 234-250.

[161] Stanford D.H and M. Hlynka (1996) "Observing general service queues before joining," *Operations Research Letters* 18, 237-245.

[162] Stenbacka, R. and M.M. Tombak (1995) "Time-based competition and the privatization of services," *The Journal of Industrial Economics* XLIII, 435-454.

[163] Stidham, S. Jr. (1978) "Socially and individually optimal control of arrivals to a $GI/M/1$ queue," *Management Science* 24, 1598-1610.

[164] Stidham, S. Jr. (1985) "Optimal control of admissions to a queueing system," *IEEE Transactions on Automated Control* AC-30, 705-713.

[165] Stidham, S. Jr. (1992) "Pricing and capacity decisions for a service facility: stability and multiple local optima," *Management Science* 38, 1121-1139.

[166] Sumita, U., Y. Masuda, and S. Yamakawa (2001) "Optimal internal pricing and capacity planning for service facility with finite buffer," *European Journal of Operational Research* 128, 192-205.

[167] Tapiero, C.S. and D. Zuckerman (1979) "Vehicle dispatching with competition," *Transportation Research B* 13B, 207-216.

[168] Tilt, B. and K.R. Balachandran (1979) "Stable and superstable customer policies with balking and priority options," *European Journal of Operational Research* 3, 485-498.

[169] Tullock, G. (1967) "The welfare cost of tariffs, monopolies, and theft," *Western Economic Journal* 5, 224-232.

[170] Vickrey, W.S. (1969) "Congestion theory and transport investment," *The American Economic Review* 59, 251-260.

[171] Walters, A.A (1968) *The Economics of Road User Charges* World Bank Staff Occasional Papers Number Five, The Johns Hopkins University Press, Baltimore.

[172] Wardrop. J.G. (1952) "Some theoretical aspects of road traffic research communication networks," *Proceedings of Industrial and Civil Engineering*, Part 2, Vol. 1, 325-378.

[173] Whang, S. (1989) "Cost allocation revisited: an optimality result," *Management Science* 35, 1264-1273.

[174] Whang, S. (1990) "Alternative mechanisms of allocating computer resources under queueing delays," *Information Systems Research* 1, 71-88.

[175] Whitt, W. (1986) "Deciding which queue to join: some counterexamples," *Operations Research* 34, 55-62.

[176] Wilson, R. (1993) *Nonlinear Pricing*, Oxford University Press.

[177] Xu, S. H. (1994) "A duality approach to admission and scheduling controls of queues," *Queueing Systems: Theory and Applications* 18, 273-300.

[178] Yechiali, U. (1971) "On optimal balking rules and toll charges in the $GI/M/1$ queue," *Operations Research* 19, 349-370.

[179] Yechiali, U. (1972) "Customers' optimal joining rules for the $GI/M/s$ queue," *Management Science* 18, 434-443.

[180] Yechiali, U., E. Altman, T. Jimenez, and R. Núñez-Queija (2000) "Queueing analysis for optimal routing with partial information."

Index

CPSIA information can be obtained at www.ICGtesting.com
Printed in the USA
LVOW10*1322240416

485090LV00003B/153/P